ARITHMETIC

To allow for the full introduction of decimalised currency
and metric units of measurement, the original version of this
book has been thoroughly revised and some completely
new chapters have been added.

The subject is developed comprehensively from its
beginnings, the elementary processes being dealt with in the
first few chapters. There follow chapters on the metric
system and on money, on 'Civil Arithmetic'—Income Tax,
Insurance, Investment—and finally on the basic arithmetic
of computers.

TEACH YOURSELF BOOKS

ARITHMETIC

Decimalised and Metricated

L. C. Pascoe, M.A. (Oxon.)

TEACH YOURSELF BOOKS
Hodder and Stoughton

First printed 1958
First printed in this form 1971
This impression 1978
New edition 1979

Copyright © 1971
L. C. Pascoe

This volume is published in the USA by David McKay
Company Inc., 750 Third Avenue, New York, NY 10017

ISBN 0 340 24285 X

Printed in Great Britain
for Hodder and Stoughton Paperbacks,
a division of Hodder and Stoughton Ltd,
Mill Road, Dunton Green, Sevenoaks, Kent
(Editorial Office: 47 Bedford Square, London WC1 3DP),
by Hazell Watson & Viney Ltd, Aylesbury, Bucks

INTRODUCTION

The original version of this book was published in 1958. The time has now come for a new look at arithmetic. Many of the principles explained in detail in the former book still apply but, with the full introduction of decimalised currency and metric units of measurement, substantial modifications are necessary. Chapters 6 and 7 are entirely new and all of the other chapters have been altered or partially rewritten to a greater or lesser extent. The result is virtually a new concept of arithmetic for the British man and woman.

The subject is developed from its beginnings, little need arising for significant amendment to the text of the first few chapters, although many of the examples are entirely new.

Chapters 6, 11 and 13 require particular care as, from the point of view of manipulation, the units of measurement may not be entirely familiar to the reader. Standard International units form the basis of measurement (Ch. 6). (*a*) For length, we have the *metre* (about 39·37 inches), its multiple, the *kilometre* (about ⅝ mile), and its submultiple, the *millimetre* (about 0·039 inch), which is very small, as a consequence of which we make widespread use of the *centimetre*, of which there are about 2·54 to the inch. (*b*) For weight, which should be called *mass*, as explained later, the standard is the *gramme*; this, being about 0·035 ounce, is too small for everyday life and we tend to utilize the *kilogramme* (about 2·2 lb) and the *tonne*, or *metric ton*, which is only 35 lb short of the former Imperial ton.

Chapters 14–16, as in the earlier book, are devoted to Civil Arithmetic, including Income Tax, Insurance and Investment. It has been necessary to reconstruct the approach to these topics, because of the evolutionary changes of the past 12 years. The final chapter, devoted to the basic arithmetic of computers, leads directly to the author's *Teach Yourself New Mathematics*, published in March 1970, which is largely a continuation of higher arithmetic and logic needed for a simple understanding of electronic calculators and computers.

Quite apart from the rethinking consequent on our new system, appreciable modifications have been necessary because of inflation. One cannot write a realistic Arithmetic without carefully assessing the impact of this factor on our lives. The author recalls his study of meteorology many years ago, when it was pointed out to him that a weather-forecaster had been known to pronounce, from his synoptic chart, that there would be brilliant sunshine all day; the unfortunate officer was brought to book by a pilot standing near him, who enquired what all the wet stuff was, which could be seen through the met. office window, descending from heaven. In the present book we take frequent observations from our window.

I should like to express my appreciation to Mr D. Bowen, for his advice on Stock Exchange procedure; to Mr C. Bolton, for helping to check on the acceptable units of measurement coming into general use; and to English Universities Press for their unfailing consideration during the preparation of this work. Above all, however, my gratitude goes to my wife, Winsome, without whose unswerving loyalty and encouragement none of my books would have had a chance of seeing the light of day.

L. C. PASCOE (1970)

Preface to the 1979 edition

Since the decimalised and metricated edition first appeared in 1970, there has been a great increase in the cost of living. Noticeable changes have also taken place in other respects, particularly with regard to taxation and investment. It has been necessary, therefore, to revise numerous examples and to modify a number of chapters, especially chapters 14, 15 and 16. The next chapter, 17, is entirely new and gives a simple introduction to Value Added Tax. It is hoped that this will prove useful to many people.

Mr R. Buckler–David has been most helpful with regard to the insurance section, as has Mr K. Wharton with chapter 17.

L.C.P.

CONTENTS

CHAPTER 1

THE BIRTH OF ARITHMETIC

Para.   *Page*   *Para.*   *Page*
1. The Origins of Arithmetic   .   1   3. Money, Weights and Measures   3
2. Development of Numerals   .   1

CHAPTER 2

ARITHMETIC—THE FOUR RULES

1. What is Arithmetic?   .   .   8   4. The Four Rules   .   .   .   9
2. Numbers   8   5. Brackets and Miscellaneous
3. Tables   8   Signs   16

CHAPTER 3

FACTORS

1. Factors   19   5. Division by Factors   .   .   21
2. Tests for Factors .   .   .   19   6. Highest Common Factor   .   22
3. Prime Factors   .   .   .   20   7. Lowest Common Multiple   .   22
4. Indices   20

CHAPTER 4

FRACTIONS

1. The Idea of a Fraction .   .   24   5. Multiplication of Fractions   .   29
2. Definition of a Fraction   .   25   6. Division of Fractions   .   .   30
3. Fractions of Standard Quanti-   7. Mixed Signs   .   .   .   31
   ties   26   8. Problems   33
4. Addition and Subtraction of
   Fractions   .   .   .   27

ix

CHAPTER 5

DECIMALS

1. Introduction . . . 35
2. Addition and Subtraction . 36
3. Multiplication and Division— Introduction . . . 37
4. Multiplication . . . 38
5. Division 39
6. Fractions and Decimals . 42

CHAPTER 6

THE METRIC SYSTEM

1. Origin of the Metric System . 44
2. The Metric System and Standard International Units . 44
3. Length 45
4. Area 47
5. Volume and Capacity . . 50
6. Mass 50

CHAPTER 7

MONEY

1. The British Coinage . . 54
2. Use of the New Coinage . 57
3. Multiplication and Division of Money 60
4. Delay in completion of Conversion from Imperial to Metric Units . . . 62
5. Practice 63
6. Unitary Method . . . 64

CHAPTER 8

RATIO, PROPORTION, AVERAGES

1. Ratio 66
2. Variation and Proportion . 68
3. Proportional Parts . . 72
4. Averages 73

CHAPTER 9

PERCENTAGE, PROFIT AND LOSS

1. Percentage . . . 76
2. Percentage Profit . . 78
3. Harder Percentage Problems. 81
4. Mixtures 85

CHAPTER 10

ARITHMETICAL GRAPHS

1. Plotting and Interpreting
   Graphs 87

2. Histograms and Frequency
   Distributions . . . 92

CHAPTER 11

MENSURATION OF RECTANGULAR FIGURES

1. The Area of a Rectangle . 97
2. Perimeter 98
3. Applications to Plane Figures 99
4. Volume of a Cuboid . . 102
5. The Box 104
6. The Area of a Triangle . . 105

7. The Area of a Trapezium . 107
8. The Volume of a Right Prism 108
9. Square Root . . . 110
10. Pythagoras' Theorem and
    Right-angled Triangles . 113

CHAPTER 12

LOGARITHMS

1. Indices 118
2. Graphical Considerations . 120
3. Logarithms . . . 120
4. The Four Rules of Logarithms 123

5. Negative Characteristics . 126
6. Harder Problems and Applications to Formulæ . . 130

CHAPTER 13

THE CIRCLE, CYLINDER, CONE AND SPHERE

1. Definitions; Properties of a
   Circle 134
2. The Cylinder . . . 137

3. Material used in making a
   Pipe 140
4. The Cone 143
5. The Sphere 147

CHAPTER 14

SIMPLE AND COMPOUND INTEREST

1. Simple Interest . . . 150
2. Inverse Problems on Simple
   Interest 151
3. Compound Interest . . 153
4. Compound Interest Formula . 155

5. Compound Interest by Logarithms 157
6. Repayment of Loans by
   Instalments . . . 158

CHAPTER 15

RATES, TAXES AND INSURANCE

1. Rates and Taxes .   .   . 160   4. Income Tax   .   .   . 162
2. Rates   160   5. Insurance 164
3. Water Rate   .   .   . 162

CHAPTER 16

INVESTMENTS AND THE STOCK EXCHANGE

1. What are Stocks and Shares? 169   6c. Contract Stamp   .   . 177
2. The Stock Exchanges .   . 170   7. Documents Involved .   . 177
3. Stocks and Shares   .   . 171   8. Miscellaneous Examples   . 179
4. Definitions of Certain Terms . 172   9. Some Types of Stocks and
5. Methods of Calculation .   . 172   Shares   .   .   . 180
6. Expenses Involved   .   . 175   10. Income Tax   .   .   . 181
6a. Stamp Duty   .   .   . 176   11. Unit Trusts   .   .   . 182
6b. Brokerage   .   .   . 176

CHAPTER 17

VALUE ADDED TAX

1. Definition and Description   . 185   2. Simple Applications of VAT   185

CHAPTER 18

COMPUTING MACHINES AND THE BINARY SCALE

1. Calculating Machines .   . 191   2. The Binary Scale .   .   . 192

ANSWERS   .   .   . 201   LOGARITHMS .   .   . 213
INDEX   .   .   . 217

TABLES

MULTIPLICATION TABLE

×	1	2	3	4	5	6	7	8	9	10
1	1	2	3	4	5	6	7	8	9	10
2	2	4	6	8	10	12	14	16	18	20
3	3	6	9	12	15	18	21	24	27	30
4	4	8	12	16	20	24	28	32	36	40
5	5	10	15	20	25	30	35	40	45	50
6	6	12	18	24	30	36	42	48	54	60
7	7	14	21	28	35	42	49	56	63	70
8	8	16	24	32	40	48	56	64	72	80
9	9	18	27	36	45	54	63	72	81	90
10	10	20	30	40	50	60	70	80	90	100

Example: $9 \times 6 = 54$

LENGTH

The standard unit of length is the *metre*, where 1 metre $\approx$ 39·37 inches.

Greek prefixes are used for units larger than the standard.

10 metres (m) = 1 dekametre (Dm)
10 dekametres = 1 hectometre
 (Dm) (hm)
10 hectometres = 1 kilometre
 (hm) (km)

Latin prefixes are used for units smaller than the standard.

1 metre (m) = 10 decimetres (dm)
1 decimetre = 10 centimetres
 (dm) (cm)
1 centimetre = 10 millimetres
 (cm) (mm)

Approximate Conversion of Length of Important Units

1 millimetre (mm) = 0·0394 inch	1 inch = 2·54 centimetres (cm)
1 centimetre (cm) = 0·3937 inch	1 foot = 30·48 centimetres (cm)
1 metre (m) = 39·37 inch (about 1·1 yd)	1 yard = 91·44 cm = 0·9144 metre (m)
1 hectometre (hm) = 109·4 yards	1 chain (=22 yards) = 20·12 metres (m)
1 kilometre (km) = 0·6214 mile	1 mile = 1·609 kilometre (km)

AREA

The standard unit of area is the *square metre*, where 1 square metre $\approx$ 1·196 square yard.

100 square metres = 1 square dekametre (Dm²), which is called the *are*
100 ares = 1 square hectometre (hm²), which is called the *hectare* (ha)
100 Ha = 1 square kilometre (km²)

Also

1 m² = 100 square decimetres (dm²)	1 cm² = 100 square millimetres (mm²)
1 dm² = 100 square centimetres (cm²)	

Approximate Conversion of Area of Important Units

1 square centimetre (cm²) = 0·155 square inch	1 square inch = 6·45 cm²
1 square metre m² = 1·196 square yards	1 square foot = 929 cm²
	1 square yard = 0·836 m²
1 are = 119·6 square yards	1 acre = 4050 m²
1 hectare (ha) = 2·47 acres	1 square mile = 2·59 km²
1 square kilometre (km²) = 0·386 square mile	

VOLUME

The standard unit of volume is the *cubic metre*, where 1 cubic metre $\approx$ 1·31 cubic yard.

The most useful units of volume are shown below

1000 cubic millimetres (mm³) = 1 cm³	1000 cubic centimetres (cm³) = 1 dm³
	1000 cubic decimetres (dm³) = 1 m³

Approximate Conversion of Volume of Important Units

1 cubic centimetre (cm³) = 0·061 cubic inch	1 cubic inch = 16·39 cm³
	1 cubic foot = 0·0283 m³
1 cubic metre (m³) = 1·31 cubic yard	1 cubic yard = 0·765 m³

MASS

This was formerly, but inaccurately, referred to as WEIGHT (see Chapter 6, pages 50–52).

The standard unit of mass is the gramme (g), where 1 g ≏ 0·0343 ounce.

10 g = 1 dekagramme (Dg)	1 g = 10 decigrammes (dg)
10 Dg = 1 hectogramme (hg)	1 dg = 10 centigrammes (cg)
10 hg = 1 kilogramme (kg)	1 cg = 10 milligrammes (mg)

Also 1 000 000 g = 1000 kg = 1 tonne (metric ton).

The practical units are the milligramme (laboratory work), the gramme (laboratory work and the kitchen), the kilogramme (the kitchen and every-day life), the tonne (heavy commodities).

Approximate Conversion of Mass of Important Units

1 mg ≏ 0·000 035 oz	1 oz ≏ 28·3 g
1 g ≏ 0·0353 oz	1 lb ≏ 0·454 kg
1 kg ≏ 2·205 lb	1 cwt ≏ 50·8 kg
1 tonne ≏ 2205 lb	1 ton ≏ 1·02 tonne

CAPACITY

1000 cm³ = 1 litre (l)   1000 l = 1 cubic metre (m³)

From the table of volume above it will be seen that 1l = 1dm³.

The *cubic centimetre* is sometimes called the *millilitre* (ml).

Approximate Conversion of Capacity of Important Units

1 litre = 1·76 pint	1 pint = 0·568 litre
1 cubic metre = 220 gallons	1 gallon = 4·543 litres

TIME

60 seconds (s) = 1 minute (min)	14 days = 1 fortnight
60 minutes (min) = 1 hour (h)	365 days = 1 year
24 hours (h) = 1 day	366 days = 1 leap year
7 days = 1 week	

Number of Days in Each Month. January 31, February 28/29, March 31, April 30, May 31, June 30, July 31, August 31, September 30, October 31, November 30, December 31.

A Leap Year is one whose date is divisible by 4, e.g. 1972, 1976, *unless* the date is divisible by 100, but not by 400. For example, 1800 and 1900 were not leap years. 2000 A.D. will be. February has 28 days in an ordinary year, 29 in each leap year.

TABLES

MONEY

100 new pence (p) = 1 pound sterling (£).
This can be written as 1p = £0·01, where 1p = 2·4d (old pence).

SPECIAL UNITS

12 units = 1 dozen (doz.) 20 units = 1 score	144 units = 1 gross

THE BIRTH OF ARITHMETIC

1. The Origins of Arithmetic

In early days counting was unknown and unnecessary. Animals were killed for food and their skins were used for clothing. A skin could be exchanged for an axe, and such a simple system of barter was adequate for the needs of primitive races. A shepherd tending his flock thousands of years ago could not have told us how many sheep he had. He could not have counted beyond three, but he would have known if one sheep were missing, for he thought of them as individuals not as numbers.

Our modern arithmetic has arisen largely because we no longer barter goods. It would not be practical in our complex way of life. We have to buy and sell, so a system of coinage has become necessary, and with it has developed a technique of calculation. People have come to live in communities, villages and towns. To provide for their needs it has proved necessary to estimate populations, and quantities of food and materials needed by them.

The development of arithmetic has been erratic, sometimes advancing by bounds and sometimes making little progress for several centuries because of difficulties in counting systems. For example, the Roman numerals were quite easily added, but were troublesome to manipulate in other ways.

2. Development of Numerals

Long ago people learned to count up to five, largely because there are five fingers on one hand. This gave rise to the seximal scale (or scale of six) in which numbers read 0, 1, 2, 3, 4, 5, 10, 11, 12. . . . This system is still the basis of counting in some parts of the world and the illustration of the Oriental abacus (or counting frame) below, which is in the author's possession, is clearly designed for use in this system. Such abaci are still in use in China and Japan, where they are called *suan pan* and *soroban* respectively. The abacus is also reputed to be still in widespread use in Russia. Great skill can be developed in the use of counting-frames, and ordinary calculations can be carried out as quickly as with pen and paper, but they hardly compete with electronic calculators!

One of the stumbling blocks to the progress of arithmetic was the

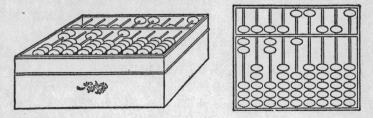

concept of zero. Men were able to understand the symbol 3, say, representing 3 horses, 3 cabbages or 3 loaves. It required more subtle thought to appreciate that a man who did not have a horse was the same as a man who owned 0 horses. Oddly enough, the author was enlightened to read in one of our national daily newspapers which specialises in letters from readers, in reply to a mother who asked whether 12 times 0 was equal to 0, as given in her son's exercise book, was correct, words approximating to 'Don't be a fool, woman. Nought times any number doesn't change it. Nought times twelve is twelve.' The author bowed his head. He had always understood that when a gift of nothing was made twelve times the recipient did not materially benefit.

Once zero was understood, great strides were made in the development of arithmetic. The decimal scale, originating in the use of all ten fingers, soon predominated, but other scales also had their day. The Babylonians counted in a scale of 60. This sexagesimal scale has survived in our measure of angle and time:

Angle: $60'' = 1'$; $60' = 1°$; $360° = 1$ complete rotation;

Time: 60 seconds = 1 minute; 60 minutes = 1 hour.

Some tropical races used a scale of 20, because they walked barefoot and used fingers and toes in counting.

A few examples of ancient and modern numerals are shown below:

Readers who have played the game of 'Mah-Jongg' will have seen the Chinese numerals on the set of pieces (or cards) called Characters.

3. Money, Weights and Measures

The ancient system of barter broke down as civilisation developed. Suppose, for example, a cobbler makes a new pair of shoes for a carpenter. The carpenter offers to make some furniture for the cobbler's home, but the cobbler already has as much as he needs. The carpenter has to find someone who will take a table in exchange for something

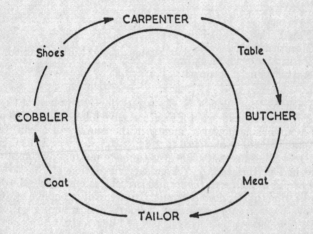

the cobbler needs. He is lucky. The butcher needs a table. He gives meat in exchange. The cobbler says he has plenty of meat. The tailor, however, needs it and presents the carpenter with a coat in exchange. This the cobbler accepts and the transaction is completed.

Now this is all very well, but the carpenter will have been put to considerable trouble to arrange this cycle of events. What can be done?

The invention of tokens, small and easily carried and accepted by all as representative of potential purchasing power, is clearly the answer, so we find the origins of currency. It has taken many strange forms. Red Indians used 'wampum', strings of beads in carefully measured lengths. In Cochin China, a strange 'coinage' existed.[1]

[1] M. E. Bowman: 'Romance in Arithmetic'.

$$\left\{\begin{array}{l}\text{40 hoes} \quad = \text{1 earthenware jar} \\ \text{7 jars} \quad\;\; = \text{1 buffalo} \\ \text{6 buffaloes} \; = \text{1 slave}\end{array}\right.$$

Each part of the world has naturally developed its own system of weights and measures, but nowadays, with so much international trade existing, it might reasonably be expected that there would be some uniformity of coinage.

This unfortunately is far from realisation. It is only necessary to consider for a moment the rates of exchange given below on 11th May 1978.

Place		
New York	$1·82	
Montreal	$2·04	Value of £1 at the city named.
Paris	8·45 fr.	
Amsterdam	4·07 fl.	

Even where the coinage is the same (for example, U.S.A. and Canada), the problem is not altogether solved, for exchange depends on the purchasing power of money in the country of origin. In the above table it will be observed that $1·82 U.S. = £1 = $2·04 Canada.

There was no necessity however to make matters as complicated as we had in this country. A cursory inspection of the tables of weights and measures of the former British system clearly shows the complications of our obsolete units. Can we really uphold a system of dry measure: 2 pints = 1 quart, 4 quarts = 1 gallon, 2 gallons = 1 peck, 4 pecks = 1 bushel, 8 bushels = 1 quarter?

The following example illustrates our elaborate form of linear measure, now superseded by the metric system:

12 inches = 1 foot (ft)	4 poles = 1 chain (ch)
3 feet = 1 yard (yd)	10 chains = 1 furlong (fur)
5½ yards = 1 pole (p)	8 furlongs = 1 mile

Consider[2]

mile	fur	ch	p	yd	ft	in
3	7	9	3	5	1	7
						2×
2)8	0	0	0	0	0	2
4	**0**	**0**	**0**	**0**	**0**	**1**

[2] Second Penguin Problems Book.

We began with all quantities of the right size. We then multiplied a quantity of less than 4 miles by 2. We divided the result by 2 and arrived at a total greater than 4 miles—or so it seems! The fault lay in our curious system of measurement. The student will no doubt verify this for himself.

Compare this with the metric system (Chapter 6). One standard unit exists for mass (gramme) and one for length (metre). Everything else is found by multiplying by powers of ten, e.g. 1 *kilo*gramme = 1000 grammes, 1 *centi*metre = $\frac{1}{100}$ metre.

Our money, also, was difficult to handle. How much easier than 12d. = 1s., 20s. = £1 is 100p = £1, even although it may be argued that the new penny is too big.

We conclude the chapter with one or two examples of calculation from other times and countries.

Ex. 1.

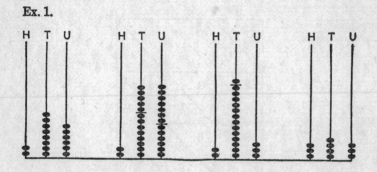

Addition with the ancient abacus. Problem: add 57 and 286. (The headings stand for hundreds (H), tens (T), units (i.e. numbers less than 10) U.)

(1) Put 2 pebbles in H column, 8 in T column, 6 in U column = 286.

(2) Add 5 in T column and 7 in U column = 57 added.

(3) Throw out 10 in U column, add 1 in T column.

(4) Throw out 10 in T column, add 1 in H column.

Answer **343** (or more probably CCCXLIII).

Ex. 2. Multiplication used by the peasants of Russia, involving the two times table and division by 2 only.

Multiply 27 by 73.

The explanation is as follows. Put the smaller number in the first column. Divide by 2 again and again ignoring remainders where present, placing the result underneath each time. Put the larger number in the second column. Double it again and again, placing the result underneath each time.

27	73
13	146
6	292
3	584
1	1168
	1971

Strike out all numbers in the second column which come opposite even numbers in the first column. Add up the rest.

Ex. 3. We conclude with an example of multiplication in the 15th century. This is particularly interesting in that it is much simpler than our present system but there is a drawback in that a diagram, called a *grid*, has to be drawn before commencing battle.

Multiply 293 by 347.

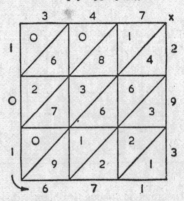

347 times 293 equals **101 671.**

Explanation. (i) Multiply each figure of the top line by each figure down the right-hand side, filling in the results as shown.

(ii) Read off the totals diagonally downwards, starting by taking the units (1) in the bottom right-hand corner, tens $4+6+6+1+9 = 26$, put down 6, carry 2 to next diagonal, and so on.

Advantages. This method never requires multiplication by more than 9. All figures are put down at once, and all multiplication is completed before addition takes place.

Disadvantages. A careful diagram is necessary and for large numbers it tends to take up a considerable amount of room. Also the diagram changes shape from number to number (e.g. 3602×27 is 4 squares by 2 squares, 293×347 (above) is 3 squares by 3 squares).

Intelligent children of about 9 years of age have been shown this method and have taken some 2 minutes or less to master it! Adults do not require so long, one hopes!

EXERCISE 1

1. Do some simple multiplication of two numbers using the methods of Ex. 2 and Ex. 3 above. Check the results by normal multiplication.

CHAPTER TWO

ARITHMETIC—THE FOUR RULES

1. What is Arithmetic?

Arithmetic is the study of the numbers 1, 2, 3, 4 . . . under various operations of which the simplest are addition, subtraction, multiplication and division. These are the so-called 'Four Rules'. Later we shall add other processes and learn to apply them to everyday life.

Speed and accuracy in simple calculations must first be mastered, and the student is advised to work as many examples as possible.

The symbols used nowadays are Arabic in origin, this system being much simpler to apply than the previous Roman method. Imagine a multiplication sum, MDCXL multiplied by CCLIX!

The word 'Arithmetic' is derived from the Greek *arithmos*, meaning number.

2. Numbers

The numbers we use are said to be in the scale of ten, that is, when a number is multiplied by 10 it is placed in the next space to the left, e.g. | 5 | 6 | times 10 becomes | 5 | 6 | 0 |.

For example, 4 3 6 5 consists of:

4 thousands 3 hundreds 6 tens and 5 units, i.e.
4000 and 300 and 60 and 5.

If we were to multiply this number by 10 we would have 4 3 6 5 0, in which the 4 stands for 4 ten-thousands, the 3 for 3 thousands, and so on.

Other scales are sometimes used and to these brief reference will be made for interest in the last chapter of this book. The binary scale (or scale of two) is of great importance in the theory of modern electronic computing machines.

3. Tables

It is necessary that the student should be fully conversant with the multiplication tables up to the 12-times table, and an abbreviated form of table is given with the tables of measures, for completeness (page xiii).

Celerity in addition and subtraction is also of value, and it is recommended when adding, say, 7, 12, 5 and 8 that instead of saying mentally '7 and 12 are 19, 19 and 5 are 24, 24 and 8 are **32**' the shorter method '7, 19, 24, **32**' be practised.

4. The Four Rules

(a) **Addition.** The symbol for addition is $+$ (plus), from the Latin, meaning more; it is placed between two numbers to be added together. Thus, $8+7$ means 'eight plus seven' or 'seven added to eight'. We can use the symbol repeatedly between numbers to be added, e.g. $8+4+9+12$.

The symbol $=$ means 'is equal to', so from above we have

$$(a) \ 8+7 = 15$$
$$(b) \ 8+4+9+12 = 33.$$

We can add either in a *row* or in a *column*, but when the numbers are large we always use the latter method.

Ex. 1. $17+5+123 = 145$.

The units are first added: $7+5+3 = 15$; put down **5**, carry 1.

The tens are next added: $1+0+2 = 3$, add 1 carried from units; put down **4**.

Hundreds column: put down **1**.

Result **145**.

Ex. 2. Add together 73, 1204 and 513.

We have

$$
\begin{array}{r}
73 \\
1204 \\
513 \\
\hline
1790 \\
\end{array}
$$

Adding the units, $3+4+3 = 10$; put down **0**, carry 1 to tens column.

Adding the tens, $7+0+1 = 8$, add 1 carried from units, giving **9**.

Adding the hundreds, $2+5 = 7$. Put down **7**.

Thousands. Put down **1**.

Result **1790**.

Great care must be taken in working in columns, to see that units appear under units, tens under tens, and so on. The units column is the guide to the lay-out.

Ex. 3. Jones has to undertake a business trip to Newcastle-upon-Tyne calling at Cambridge, Nottingham, Sheffield and Leeds on the way. He starts at London. How far has he to go?

From a map:

	kilometres
London–Cambridge	82
Cambridge–Nottingham	134
Nottingham–Sheffield	59
Sheffield–Leeds	52
Leeds–Newcastle	147
	474
	2 2

The carrying figure from the units column is placed under the tens column, and then added into the tens column. The carrying figure from the tens column is placed under the hundreds column, and then added into the hundreds column.

For certain purposes, where accuracy is of prime importance, the following system of checking results can be used.

Ex. 4. In an examination, Smith, Brown, Jones, and Robinson had 5 questions to answer, each carrying 20 marks. The following table gives the results:

Name	Question No.					Max. 100
	1	2	3	4	5	
Smith	16	9	18	20	12	75
Brown	8	14	3	16	15	56
Jones	15	17	5	17	14	68
Robinson	12	14	2	13	14	55
	51	54	28	66	55	**254**

Although the results shown in the last column may give all that is required, by adding the total marks gained in each *question* and then summing the results in the bottom row, and also in the last column, and seeing that the total (254) is the same both ways, we obtain a valuable check on accuracy. It is also interesting to note that this table shows clearly that there was some weakness in answering question 3, where a total of only 28 marks was gained by all candidates out of a possible 80.

EXERCISE 1

1. Add together 17, 12, 6 and 8.

2. Find the value of $3+7+19+5+8$.

3. Add   (a) 124   (b) 1496   (c) 4009

              37   302   378

          209   617   9152

4. Six workmen, Williams, Brown, Andrews, Jenkins, Smart and Thompson, receive varying wages from a factory according to the amount of overtime done. In four successive weeks, they are paid as follows:

Williams   £72, 66, 78, 75
Brown   £66, 75, 81, 66
Andrews   £75, 78, 75, 84
Jenkins   £72, 63, 78, 75
Smart   £57, 60, 66, 60
Thompson £81, 90, 78, 93

Find how much each receives, and the total wages bill of the factory for the 4 weeks. Check your working as in Ex. 4 above.

(b) Subtraction. The symbol for subtraction is — (minus), from the Latin, meaning less; it is placed between two numbers, when the second is to be taken away from the first. For example, $12-9$ means 'twelve minus nine' or 'nine taken away from twelve'; we get $12-9 = 3$.

Ex. 5. $147-59 = \mathbf{88}$.

Here we cannot take 9 from 7, so we borrow from the tens column and read as follows: 17 take 9 gives **8**, put down; add back 1 to the 5 to be taken away, to make up for the 1 borrowed. Read 14 take 6 gives **8**, and put this in the tens column of the answer.

Ex. 6. In a mixed school of 1,847 children, there are 792 girls. How many boys are there?

18^147
7_192
1055

By putting 1 borrowed from hundreds column next to the 4, making 14 tens and repaying it by putting it under the 7 in the hundreds column, we proceed as follows.

7 take 2 gives **5**   ⎫
14 take 9 gives **5**   ⎬ Result **1055**
8 take $(7+1)$ gives **0** ⎬
1 take 0 gives **1**   ⎭

We can use $+$ and $-$ signs in the same line, indicating amounts added and subtracted.

Ex. 7. Williams had 54 Savings Certificates. He bought 12 more but later sold 30 of them to pay for a new radio. How many had he left

$$54+12-30 = 66-30$$
$$= \mathbf{36}$$

(i) We *add* the numbers to be *added* first.

(ii) We *add* the numbers to be *taken away*. (In the above example there was only one such number.)

(iii) We *take* the second result from the *first* result.

Ex. 8. $43+8-24-16 = 51-40$
$$= \mathbf{11}.$$

Ex. 9. In an election there were 3 candidates, Messrs. Green, Jones and Black. Mr. Green secured 12 542 votes, Mr. Jones had 8577 votes. If the total number of votes cast was 31 790, how many people voted for Mr. Black?

In this problem it is better to work as follows because of the large numbers.

12 542 $(+)$	31 790 $(-)$
8 577	21 119
21 119	**10 671**

Mr. Black secured 10 671 votes.

Here we have added the total of votes not cast for Mr. Black, and subtracted the result from all the votes cast.

EXERCISE 2

1. Subtract (*a*) 41 from 57   (*b*) 38 from 83.
2. Find the value of (*a*) $271-141$ (*b*) $328-279$.
3. A school has 54 boys under 12, 57 under 13 but over 12, 62 under 14 but over 13, 53 under 15 but over 14, and 60 under 16 but over 15. Altogether there are 352 boys in the school. How many are over 16?
4. Fill in the blanks with $+$ or $-$ in the right places.

    (*a*) 8   4   17 $= 29$
    (*b*) 25   8   3 $= 20$
    (*c*) 36   15   24 $= 27$
    (*d*) 17   5 $= 37$ 25.

(c) **Multiplication.** When we have a number added to itself several times, we shorten the process considerably by multiplication. If seven rows of strawberry plants are laid out in a garden with 12 plants in each row, we could either add:

$12+12+12+12+12+12+12 = 84$ (7 lots of 12)

or

$7+7+7+7+7+7+7+7+7+7+7+7 = 84$ (12 lots of 7).

The working is greatly reduced by writing either $7 \times 12 = 84$ or $12 \times 7 = 84$, and reading as 'seven times twelve equals eighty-four' or 'twelve times seven equals eighty-four'. The symbol $\times$ means 'multiplied by' or 'times'.

A set of multiplication tables is given on page xiii, as mentioned earlier, and these should be memorised.

Ex. 10. What is the cost of a fleet of a dozen cars at £2827 each?

$$2827 \times 12$$
$$\underline{12}$$
$$\mathbf{33924}$$
$${}_{9\,3}{}^{8}$$

Total cost **£33924**

The carrying figures can be placed under the correct columns as in addition.

$12 \times 7 = 84$; put down **4**, carry 8.
$12 \times 2 = 24$; add 8, making 32, put down **2**, carry 3.
$12 \times 8 = 96$; add 3, making 99, put down **9**, carry 9.
$12 \times 2 = 24$; add 9, making 33, put down **33**.

For numbers greater than 12, long multiplication is employed. We then multiply by the units, then the tens and so on.

Ex. 11. Multiply 742 by 397.

We can think of it as 742×7 added to 742×90 added to 742×300. Now multiplying by 10 is achieved by adding a nought to the number.

$\therefore 742 \times 90 = 7420 \times 9$; $742 \times 300 = 74\,200 \times 3$.

$$742 \times 397$$
$$\underline{397}$$

	5 194	$(= 742 \times 7)$
(i)	66 78	$(= 7420 \times 9)$
(ii)	222 6	$(= 74\,200 \times 3)$

$$\mathbf{294\ 574}$$

We do not actually write the noughts indicated but move the figures along one place to the left, so that units become tens and so on in line (i) above; we move the figures two places to the left in line (ii) above, so that the units become hundreds.

1. Multiply (a) 17 by 5   (b) 23 by 7   (c) 168 by 9   (d) 214 by 11.

2. Find the value of (a) 34×12 (b) 417×6 (c) 3925×8 (d) 9579×12.

3. Evaluate (a) 317×16 (b) 624×19 (c) 529×73 (d) 8912×106.

4. Find the value of $23 \times 7 \times 9 \times 4$.
   [*Hint.* $23 \times 7 \times 9 \times 4 = 161 \times 9 \times 4$, and so on.]

5. Evaluate $3401 \times 17 \times 5$.

(d) **Division.** Division is the process of sharing. The sign for division is $\div$; for example, $12 \div 3$ means 'twelve divided by three' or 'if twelve were divided into three equal groups, how many would there be in each group'.

Ex. 12. $132 \div 12 = 11$ (because 12 lots of 11 make 132).

Ex. 13. Divide 4988 by 4.

We proceed as follows:

$$\begin{array}{r} {\scriptstyle 1\,2} \\ 4)\overline{4988} \\ \hline \mathbf{1247} \end{array}$$

The steps are 4 into 4 divides **1**, put down **1**; 4 into 9 divides **2**, put down, and carry 1 to next column; 1 hundred$+8$ tens $= 18$ tens (i.e. we place the 1 in front of the 8); 4 into 18 divides **4**, put down, and carry 2 to next column; 2 tens$+8$ units $= 28$ units; 4 into 28 divides **7**, put down. Result **1247**.

This process is called *short division*.

When the number by which we divide (called the *divisor*) is greater than twelve we usually resort to *long division*, shown below.

Ex. 14. Find the value of $1161 \div 43$.

$$\begin{array}{r} \mathbf{27} \\ 43)\overline{1161} \\ 86 \\ \hline 301 \\ 301 \\ \hline \cdots \end{array} \qquad \text{Result } \mathbf{27}.$$

The number by which we divide (43 in this case) is the *divisor*.

The number into which we divide (1161) is the *dividend*.

The number obtained (27) is the *quotient*.

The process is as follows: 43 will not divide into 1 or 11 but it will go into 116. We try the largest number of times it will go, i.e. **2**. Put this in the quotient space and write the result of 43×2, i.e. 86, under the 116 and subtract. We get 3 (hundreds) 0 (tens). Bring down the 1 (unit), giving 301, 43 goes into 301 *seven* times exactly. Enter 7 in the quotient next to the 2, obtaining 27.

It often happens that division cannot be carried out exactly.

Ex. 15. A car runs 13 kilometres on 1 litre of petrol. How many litres are needed for a journey of 387 kilometres?

$$
\begin{array}{r}
29 \\
13)\overline{387} \\
26 \\
\hline
127 \\
117 \\
\hline
10 \leftarrow \text{Remainder}
\end{array}
$$

The result is **29, remainder 10**, i.e. $29\frac{10}{13}$.

Here we see that 29 litres are needed, plus 10 parts out of 13 parts (i.e. nearly 30 litres).

In order to allow for petrol wastage in traffic queues and other hazards, the wise motorist would probably allow himself 35 litres.

Another method of division for many numbers greater than 12 is shown in the chapter on Factors.

A little thought will show that the above examples are based on the principle

$$\text{Dividend} = \text{Divisor} \times \text{Quotient} + \text{Remainder}.$$

(The multiplication sign applies only to the quantity on either side of it, not to the Remainder.)

Ex. 16. What number when divided by 7 gives a quotient of 9 and a remainder 4?

$$
\begin{aligned}
\text{Dividend} &= 7 \times 9 + 4 \\
&= 63 + 4 \\
&= \mathbf{67}.
\end{aligned}
$$

EXERCISE 4

1. Divide (*a*) 3177 by 9 (*b*) 2706 by 11 (*c*) 298 554 by 37.

2. Express 37 206 inches in metres (take 1 metre as approximately 39 in).

3. 1102 people attend a lecture and are seated on chairs 29 in a row. How many rows are there?

4. What is the smallest number above 2554 which is exactly divisible by 7?

5. The circumference of a bicycle wheel is 208 cm. How many times will it rotate in travelling 100 metres?

5. Brackets and Miscellaneous Signs

Brackets are used when a group of numbers is to be treated as a single number.

Ex. 17.
$$3 \times (4+5) = 3 \times 9$$
$$= 27.$$

Ex. 18.
$$42 - (17+3) = 42 - 20$$
$$= 22.$$

The part inside the bracket is worked out first. More than one pair of brackets may be used, and in this case the innermost bracket is evaluated first.

Ex. 19.
$$6 \times (3+4) - 4 \times (5-2) = 6 \times 7 - 4 \times 3$$
$$= 42 - 12$$
$$= 30.$$

Ex. 20.
$$63 - \{7 + 2 \times (8-5)\} = 63 - \{7 + 2 \times 3\}$$
$$= 63 - \{7 + 6\}$$
$$= 63 - 13$$
$$= 50.$$

When using different signs together as above, the word BODMAS is helpful, indicating:

1	B	Brackets
2	O	Of
3	D	Division
4	M	Multiplication
5	A	Addition
6	S	Subtraction

This gives the order in which the working must be done, starting with 1 (Brackets). 'Of' means 'multiply', but is not often used, e.g. 5 of 7 = 35. Unlike multiplication, when it is used, 'of' *takes priority over division*.

1. Deal with inside of Brackets.
2. Work out Of (if any).
3. Work out Division.
4. Work out Multiplication.
5 and 6. Work out Addition and Subtraction as previously explained.

Ex. 21. Simplify 3 of $(12-7)+56\div(9-5)$.

The expression	$= 3$ of $5+56\div4$	Rule 1
	$= \quad 15 \quad +56\div4$	Rule 2
	$= \quad 15 \quad +14$	Rule 3
	$= \mathbf{29}.$	Rule 5

EXERCISE 5

1. Add together (a) $24+6+9+17$ (b) $23+5+67+8$ (c) $424+137+19$.

2. Add (a) 21 (b) 231 (c) 25 (d) 31 725
   43   79   2077   9 406
   162   928   165   784
                        4821   23 209

3. Subtract (a) 147 from 308 (b) 49 from 2116.
4. Find the value of (a) $627-239$ (b) $4006-1879$.
5. Add 243, 1629, 17 and 405, and subtract 298 from the result.
6. Find the value of $2971+3046-2188$.
7. Multiply (a) 217 by 9 (b) 46 by 17
   (c) 837 by 29 (d) 251 by 367.
8. Find the value of (a) 1247×8 (b) 304×73
   (c) 295×174 (d) 3528×208.
9. Find the value of (a) $108\div9$ (b) $259\div7$
   (c) $1465\div5$ (d) $25\,073\div6$.
10. Divide (a) 2148 by 43 (b) 2242 by 59
    (c) 3207 by 88 (d) 20 173 by 133.
11. Simplify (a) $61+(23-15)\times2$
    (b) $30-(17+4)\div3$
    (c) $4+2\times\{3+4\times(8-3)-6\}$
    (d) 3 of $(5+2)-2\times(3+7)\div5$.
12. Simplify (a) $6-12\div4+3\times8$
    (b) $73-\{65\times2-(14\times5-8)\}$.
13. A book has 195 pages. There are about 284 words on each page. How many words are there in the book?
14. Queen Elizabeth I came to the throne in 1558 and died in 1603. How long did she reign?
15. A train consists of an engine of mass 95 metric tons; a tender of mass 40 metric tons and 12 carriages each of mass 32 metric tons. What is the total mass of the train?
16. 10 000 cabbages are planted, 85 in a row. How many rows are there? (The last row is incomplete.)
17. How many screws are there in 16 boxes each containing a gross?
18. George Orwell has written a book called '1984'. How many hours will there be in that year?

19. Smith earns £92 a week. How much does he get in a year (taken to be 52 weeks)?

20. A train starts on a journey of 760 kilometres across Europe. After travelling for 17 hours at an average speed of 41 kilometres per hour, how much farther has it to go?

21. How many seconds are there in a day?

22. In working out the multiplication sum 1723 × 27 a boy misreads the first number as 1732. What is the error in his answer?

23. How many numbers between 100 and 1000 are:
> (a) not divisible by 2
> (b) not divisible by 3
> (c) not divisible by either 2 or 3?

24. A motorist pays £144 a year for tax and insurance on his car. Repairs cost him £180 a year. He averages 270 kilometres travel a week in the car, and he uses one litre of petrol every 12 kilometres. He needs 1 litre of oil every 500 kilometres. If petrol costs 17p a litre and oil is 75p a litre, how much does it cost him to run his car for a year (of 52 weeks), allowing £250 for depreciation over the year? (Give the answer to the nearest £1.)

CHAPTER THREE

FACTORS

1. Factors

Consider $12 = 3 \times 4$; 3 and 4 are said to be *factors* of 12. 12 is also equal to 6×2 and to 1×12, so all the numbers, 1, 2, 3, 4, 6, 12 are factors of 12. 12 is said to be a *multiple* of 3, and of 6, and in fact of any of these numbers. We are thus led to the definition of factor.

A number x is a factor of another number y, if x divides exactly into y (or if y is a multiple of x).

Many numbers have no factors other than themselves and *unity* (i.e. one), e.g. $7 = 1 \times 7$, but 7 has no other factors; 13 possesses only factors 1 and 13. These numbers are called *prime numbers*. The simplest ones are 1, 2, 3, 5, 7, 11, 13, 17, 19, 23, 29. . . .

For generations attempts have been made to obtain a formula to give prime numbers only, but they have all failed. The method below, known as the 'Sieve of Eratosthenes' will enmesh all numbers up to the largest written down.

Strike out all multiples of 2.
Then all multiples of 3 left.
Then all multiples of 5 left.
Then all multiples of 7 left (there are none left in this diagram) and so on.

1	2	3	4	5	6	7	8
9	10	11	12	13	14	15	16
17	18	19	20	21	22	23	24
25	26	27	28	29	30	31	32
33	34	35	36	37	38	39	40

etc.

The numbers 2, 3, 5, 7 are not themselves deleted. The numbers remaining are the prime numbers.

2. Tests for Factors

The following tests are very useful and are worth memorising:
(1) A number is divisible by **2**, if the last figure is even (e.g. 1658).
(2) A number is divisible by **3**, if the sum of the figures is divisible

by 3 (e.g. 4251 is divisible by 3, because $4+2+5+1 = 12$, which is divisible by 3).

(3) A number is divisible by **4**, if the number formed by the last 2 figures is divisible by 4 (e.g. 30 528 is divisible by 4, because 28 is divisible by 4).

(4) A number is divisible by **5**, if it ends in 0 or 5 (e.g. 20 730 and 1945 are both divisible by 5).

(5) A number is divisible by **6** if it satisfies (1) and (2) above (e.g. 21 354 is divisible by 2 and by 3, therefore by $2 \times 3 = 6$).

(6) A number is divisible by **8** if the last three figures are divisible by 8 (e.g. 217 584 is divisible by 8, because 584 is divisible by 8).

(7) A number is divisible by **9**, if the sum of the figures is divisible by 9 (e.g. 8352 is divisible by 9, because $8+3+5+2 = 18$ is divisible by 9).

(8) A number is divisible by **10**, **100**, **1000**, . . . if it ends in 0, 00, 000 . . . (e.g. 7920 is divisible by 10; 304 000 is divisible by 1000).

Tests (1), (3) and (6) above are similar in nature.

Tests (2) and (7) are related; also test (4) can be associated with test (8).

Tests exist for some other factors, but they are more complicated and would be out of place here.[3]

3. Prime Factors

We have seen that $12 = 4 \times 3$. Now 3 is a prime number but 4 is not, but $4 = 2 \times 2$, and 2 is prime.

$\therefore$ **12 $= 2 \times 2 \times 3$**, and the number 12 is said to be expressed in its prime factors. They are written in ascending order of magnitude.

Ex. 1. Express 2565 in prime factors.
$$2565 = 5 \times 513 = 5 \times 9 \times 57$$
$$= 5 \times 9 \times 3 \times 19 \text{ (using the tests above)}$$
$$= \mathbf{3 \times 3 \times 3 \times 5 \times 19.}$$

4. Indices

The writing of prime factors is considerably reduced if indices are used.

In $2 \times 2 \times 2 = 2^3$, 3 is the *index*. It means that 2 is written down three times and the results are multiplied together. For example, $4^5 = 4 \times 4 \times 4 \times 4 \times 4$; $5^4 = 5 \times 5 \times 5 \times 5$. Notice that these are not the same number.

---

[3] One such test is based on the fact that $7 \times 11 \times 13 = 1001$. The test is given in detail in *New Mathematics*, Teach Yourself Books, by the present author, on pp. 83–84.

Ex. 2. $2565 = 3 \times 3 \times 3 \times 5 \times 19$ from Ex. 1 above
$= 3^3 \times 5 \times 19.$

Ex. 3. $7920 = 9 \times 880$
$= 9 \times 11 \times 80 = 9 \times 11 \times 8 \times 10$
$= 2 \times 2 \times 2 \times 2 \times 3 \times 3 \times 5 \times 11$
$= 2^4 \times 3^2 \times 5 \times 11.$

It is usual to give the answer with factors arranged in ascending order of magnitude.

5. Division by Factors

In Chapter One we made reference to a method which can be used quite often for division by a number greater than 12. This applies when the divisor can be put into factors which are less than or equal to 12. We do not necessarily resolve into prime factors in this case, as this may lead to unnecessary working.

Examples of such divisors are $18 = 9 \times 2$, $49 = 7 \times 7$, $154 = 2 \times 7 \times 11$.

Ex. 4. Find the value of $1449 \div 63$.

$63 = 7 \times 9$

$$7\overline{)1449}$$
$$9\overline{)\ 207}$$
$$23$$

Result **23.**

We divide by one factor, say, 7 as above. Divide the quotient 207 by the other factor 9.

Ex. 5. Divide 1415 by 63.

$63 = 7 \times 9$ as before

$$7\overline{)1415}$$
$$9\overline{)\ 202}\ \text{remainder 1}$$
$$22\ \text{remainder 4}$$

$\therefore 1415 \div 63 = $ **22, remainder 29** (or $22\frac{29}{63}$). The remainder is not obvious.

The second remainder 4 was found **after** dividing by 7.

$\therefore$ The total remainder was $4 \times 7 + 1 = 29$, on adding in the first remainder.

The method of division by factors is often used in monetary calculations, as will appear later.

EXERCISE 1

1. Using the tests for factors find which of the numbers 2, 3, 4, 5, 6 are factors of:

(a) 418   (b) 318   (c) 5010   (d) 30 060   (e) 11 945.

2. Using the tests for factors find which of the following numbers are divisible by *both* 8 and 9:

      (*a*) 3672   (*b*) 20 592   (*c*) 7956.

3. Resolve into prime factors without indices:

      (*a*) 18   (*b*) 60   (*c*) 924   (*d*) 3003.

4. Resolve into prime factors using indices:

      (*a*) 48   (*b*) 441   (*c*) 5808   (*d*) 21 560.

5. Find, using division by factors, the value of:

      (*a*) $1505 \div 35$ (*b*) $6816 \div 96$

      (*c*) $2709 \div 77$ (*d*) $9148 \div 27$.

6. Highest Common Factor (H.C.F.)

A *common factor* of two (or more) numbers is a factor which occurs in both of them (or all of them).

The *highest common factor* of a group of numbers is the largest number which will divide into all of them. The H.C.F. of 6 and 9 is 3, because $6 = 2 \times 3$, and $9 = 3 \times 3$, and each contains a 3.

Consider
$$\begin{cases} 60 = 2 \times 2 \times 3 \times 5 \\ \qquad\qquad\quad \downarrow\;\; \downarrow \\ 225 = \qquad 3 \times 3 \times 5 \times 5 \\ \qquad\qquad\quad \downarrow\;\; \downarrow \\ 210 = 2 \times \quad 3 \times 5 \quad \times 7 \end{cases}$$

All these numbers have common factors 3 and 5 but no others.

Therefore their H.C.F. is $3 \times 5 = \mathbf{15}$.

Ex. 6. Find the H.C.F. of 90 and 126.

$$90 = 2 \times 3 \times 3 \times 5$$
$$126 = 2 \times 3 \times 3 \times 7$$

$\therefore$ H.C.F. $= 2 \times 3 \times 3 = \mathbf{18}$. (The common factor 3 occurs twice.)

7. Lowest Common Multiple (L.C.M.)

The L.C.M. of two or more numbers is the smallest number into which they will divide exactly. To put it another way, the L.C.M. of a set of numbers is the smallest number which contains each member of the set as a factor.

Finding the L.C.M. of a set of numbers is very important when simplifying *fractions*. (See Chapter 4.)

Consider
$$\begin{aligned} 42 &= 2 \times 3 \times 7 &&= 2 \times 3 \times 7 \\ 24 &= 2 \times 2 \times 2 \times 3 &&= 2^3 \times 3 \\ 28 &= 2 \times 2 \times 7 &&= 2^2 \times 7 \end{aligned}$$

2, 2^2, 2^3 will all divide into 2^3 but will not *all* divide into any smaller number exactly; also 3 and 7 will divide into 3×7, so the L.C.M. of this set of numbers 24, 28, 42 is $2^3 \times 3 \times 7 = \mathbf{168}$.

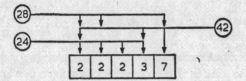

The diagram illustrates the necessity for all the factors $2 \times 2 \times 2 \times 3 \times 7$.

Exercise 2

1. Find the H.C.F. of the following sets of numbers:

    (a) 42 and 63   (b) 48, 64 and 88
    (c) 132, 154 and 242   (d) 72, 108 and 792.

2. Find the L.C.M. of the following:

    (a) 8, 12   (b) 3, 4, 8   (c) 12, 27, 36
    (d) 55, 88, 121   (e) 4, 5, 6, 10   (f) 84, 63, 42, 36
                 (g) 35, 56, 80.

3. Find the smallest number into which 30, 40 and 66 divide exactly.

4. What is the smallest sum of money which can be counted out into exact multiples of 25p, 30p and 40p?

5. Square tiles are to be used to cover a wall 12 m long to a height of 175 cm exactly. What is the largest possible size of tile? Find how many tiles are needed. (See Chapter 11 on areas. Also note that 100 cm = 1 m.)

6. Three bells are rung at intervals of 6, 8 and 9 seconds respectively. If they start together, how long is it before they are again ringing together?

7. A set of counters is required for a game in which any number of players up to ten may take part. What is the smallest number of counters required, if all the counters must be used on each occasion?

8. How many numbers between 1 and 1000 contain all the numbers 2, 3, 4, 5, 6 as factors?

CHAPTER FOUR

FRACTIONS

1. The Idea of a Fraction

So far the work in this book has been confined to whole numbers, or *integers*, as they are called. We now consider how to deal with parts of whole quantities, called *fractions*.

Suppose we buy wine in a half-litre or in a quarter-litre bottle. The quantities can be written as $\frac{1}{2}$ l or as $\frac{1}{4}$ l respectively, indicating that they are one-half ($\frac{1}{2}$) and one-quarter ($\frac{1}{4}$) of a litre. The top, called the *numerator*, tells us how many parts we are taking. The bottom, called the *denominator*, tells us how many parts would make a whole one.

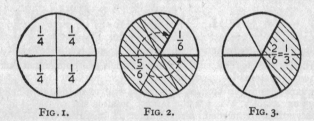

Fig. 1.   Fig. 2.   Fig. 3.

In fig. 1 we see that $\frac{1}{4}+\frac{1}{4}+\frac{1}{4}+\frac{1}{4} = \frac{4}{4}$ (4 quarters) $= 1$ (whole).

Quantities like $\frac{1}{4}$, $\frac{2}{5}$, $\frac{7}{12}$ are called *vulgar* (or *common*) fractions. The quantity $\frac{2}{5}$ means we are taking 2 parts out of 5 parts of a quantity. Consider fig. 2. Suppose it is a cake divided into six equal parts as shown. Each part is $\frac{1}{6}$ of the whole cake. The shaded area is $\frac{5}{6}$ of the whole.

Also

$$\frac{1}{6}+\frac{5}{6} = \frac{1+5}{6} = \frac{6}{6} = 1 \text{ (whole cake, in this case).}$$

What happens if the numerator and denominator have a common factor?

In fig. 3 we have divided our cake again into 6 parts but have now taken 2 of them (shaded). Suppose, however, we had divided our cake into 3 parts, as indicated by the heavier lines in fig. 3, we would have had the same shaded area as $\frac{1}{3}$ of our cake. $\therefore \frac{2}{6} = \frac{1}{3}$. This leads us to conjecture that we can cancel out common factors in numerator and denominator, e.g.

$$\frac{24}{28} = \frac{\overset{6}{\cancel{24}}}{\underset{7}{\cancel{28}}} = \frac{6}{7} \text{ (on dividing out top and bottom by 4)}.$$

The fraction is then said to be 'in its lowest terms'.

We can reverse the idea and say, for example, that $\frac{1}{3} = \frac{2}{6} = \frac{3}{9} = \ldots$

2. Definition of a Fraction

A fraction is defined as the quotient (i.e. the ratio) of two quantities.

$$\text{Fraction} = \frac{\text{Numerator}}{\text{Denominator}}.$$

A fraction like $\frac{6}{7}$ (written in words 'six-sevenths'), in which the numerator is smaller than the denominator, is called a *proper fraction*. One like $\frac{17}{12}$, in which the numerator is greater than the denominator, is called an *improper fraction*. The latter can always be reduced to a whole number and a proper fraction, known as a *mixed number*.

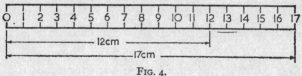

FIG. 4.

Consider fig. 4 which illustrates part of a 25 cm rule, measured in centimetres. The lengths shown are 12 cm and 17 cm. We can say the shorter length is $\frac{12}{17}$ of the longer length *or* the longer is $\frac{17}{12}$ of the shorter, i.e. the longer is

$$\frac{12+5}{12} = 1 + \tfrac{5}{12} = 1\tfrac{5}{12} \text{ of the shorter.}$$

We do not, in practice, write down all these steps. For example, $\frac{23}{5} = 4\frac{3}{5}$ is written down at once, for $23 \div 5 = 4 + \text{remainder } 3$ (that is four and three-fifths).

3. Fractions of Standard Quantities

In section 1 above we saw that we could multiply the top and bottom of a fraction by the same number without altering the result; for example

$$\tfrac{2}{3} = \tfrac{2}{3} \times \tfrac{7}{7} = \tfrac{14}{21}.$$

The reason is that $\tfrac{7}{7} = 1$, so we have multiplied only by 1. We extend the idea below, and the method is important. It will be used, for example, when working long multiplication by practice.

Ex. 1. Find the value of $\tfrac{1}{5}$ of £1.

$$\tfrac{1}{5} \times \text{£}1 = \tfrac{1}{5} \times 100\text{p} = \mathbf{20p.} \quad \left(\dfrac{1}{5} \times 100 = \dfrac{\overset{20}{\cancel{100}}}{\underset{1}{\cancel{5}}} = 20. \right)$$

Ex. 2. Find the value of $\tfrac{3}{8}$ of 1 kg.

$$\text{We have } 1 \text{ kg} = 1000\text{g}$$
$$\therefore \tfrac{3}{8} \times 1000 \text{ g} = 3 \times 125 \text{ g}$$
$$= \mathbf{375 \text{ g.}}$$

Ex. 3. What fraction is 4 m 75 cm of 7 m 25 cm?

We have 1 m = 100 cm

$$\therefore \quad \dfrac{4 \text{ m } 75 \text{ cm}}{7 \text{ m } 25 \text{ cm}} = \dfrac{475 \text{ cm}}{725 \text{ cm}} = \dfrac{\overset{95}{\cancel{475}}}{\underset{145}{\cancel{725}}} = \mathbf{\dfrac{19}{29}.}$$

Ex. 4. What fraction is $7\tfrac{1}{2}$p of £1?

$$\dfrac{7\tfrac{1}{2}\text{p}}{\text{£}1} = \dfrac{2 \times 7\tfrac{1}{2}\text{p}}{2 \times \text{£}1} = \dfrac{15\text{p}}{\text{£}2} = \dfrac{15\text{p}}{200\text{p}} = \dfrac{\overset{3}{\cancel{15}}}{\underset{40}{\cancel{200}}} = \mathbf{\dfrac{3}{40}.}$$

Ex. 5. What fraction is 15 h 45 min of 1 day?

$$\dfrac{15 \text{ h } 45 \text{ min}}{1 \text{ day}} = \dfrac{15\tfrac{3}{4} \text{ h}}{24 \text{ h}} = \dfrac{63 \text{ h}}{24 \times 4 \text{ h}} = \mathbf{\dfrac{21}{32}.}$$

It will be noticed in examples 3 to 5 that the fraction may not be worked out until *the same units are used top and bottom.*

EXERCISE 1[4]

Find the value of:

1. $\tfrac{2}{3}$rds of 1 day　　2. Four-fifths of 1 h　　3. $\tfrac{3}{8} \times 1$ litre

[4] The quantities used here all appear in the tables of money and measures, on pages xiii to xvi.

4. $\frac{7}{10}$ of 2 min   **5.** $\frac{4}{5}$ of 8 min   **6.** $\frac{7}{8} \times £1$

Find what fraction:

7. 35 kg is of 98 kg   **8.** 4 h is of 1 day   **9.** $17\frac{1}{2}$p is of £7

10. 13 h 20 min is of 1 day   **11.** 3 days 21 h 20 min is of 1 week.

4. Addition and Subtraction of Fractions

When two fractions have the same *denominator*, addition or subtraction is simple. For example, $\frac{1}{5} + \frac{2}{5} = \frac{3}{5}$.

What happens, however, when we wish to handle fractions with different denominators? Suppose we wish to find $\frac{2}{3} - \frac{1}{4}$. We find the L.C.M. of the denominator (see Chapter 2), in this case 12, and express the fractions in terms of denominator 12.

$$\frac{2}{3} = \frac{8}{12}; \quad \frac{1}{4} = \frac{3}{12}$$

$$\therefore \frac{2}{3} - \frac{1}{4} = \frac{8}{12} - \frac{3}{12} = \frac{8-3}{12} = \frac{5}{12}.$$

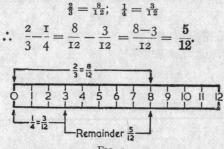

Fig. 5.

The illustration makes the process clear.

Ex. 6. Simplify $\frac{2}{5} + \frac{1}{2} - \frac{5}{6}$.

L.C.M. of denominator $= 30$

$$\therefore \frac{2}{5} + \frac{1}{2} - \frac{5}{6} = \frac{12 + 15 - 25}{30} = \frac{27 - 25}{30}$$

$$= \frac{2}{30} = \frac{1}{15}.$$

The explanation is as follows:

$$\frac{1}{5} = \frac{6}{30} \quad \therefore \frac{2}{5} = \frac{2 \times 6}{30} = \frac{12}{30}.$$

$$\frac{1}{2} = \frac{15}{30}$$

$$\frac{1}{6} = \frac{5}{30} \quad \therefore \frac{5}{6} = \frac{5 \times 5}{30} = \frac{25}{30};$$

or simply $\dfrac{2}{5}+\dfrac{1}{2}-\dfrac{5}{6} = \dfrac{2\times6+15-5\times5}{30}.$)

The student should endeavour to work this part of the question mentally as soon as possible, but for the first few examples it is easier and safer to put in the extra step indicated at the end of the bracket.

Ex. 7. Simplify $4\frac{2}{3}-1\frac{1}{4}$.
$$4\frac{2}{3}-1\frac{1}{4} = (4-1)+(\tfrac{2}{3}-\tfrac{1}{4})$$
$$= 3+\dfrac{8-3}{12} = 3\tfrac{5}{12}.$$

Note that in *addition* and *subtraction* we deal with whole numbers separately.

Ex. 8. Simplify $3\frac{1}{2}-1\frac{3}{5}$.
$$3\frac{1}{2}-1\frac{3}{5} = (3-1)+(\tfrac{1}{2}-\tfrac{3}{5})$$

$$= 2+\dfrac{5-6}{10}$$

$$= 1+\dfrac{10+5-6}{10}$$

$$= 1\tfrac{9}{10}.$$

(To make the fractional part positive we borrow 1, i.e. $\tfrac{10}{10}$, from the whole number.)

Ex. 9. Simplify $4\frac{1}{6}-2\frac{4}{5}+\frac{4}{3}$.
$$4\frac{1}{6}-2\frac{4}{5}+\tfrac{4}{3} = 4\tfrac{1}{6}-2\tfrac{4}{5}+1\tfrac{1}{3}$$
$$= (4-2+1)+(\tfrac{1}{6}-\tfrac{4}{5}+\tfrac{1}{3})$$
$$= 3+(\tfrac{1}{6}-\tfrac{4}{5}+\tfrac{1}{3})$$
$$- 3+\dfrac{5-24+10}{30} = 3+\dfrac{15-24}{30}$$
$$= 2+\dfrac{30+15-24}{30} = 2\tfrac{21}{30} = 2\tfrac{7}{10}.$$

Ex. 10. Mary was offered the option of $\frac{7}{9}$ or $\frac{17}{20}$ of a prize of 75p. Which is better?

This kind of problem leads to placing fractions in order of magnitude.

L.C.M. of 9 and 20 is 180.

$$\therefore \tfrac{7}{9} = \tfrac{140}{180}; \quad \tfrac{17}{20} = \tfrac{153}{180}$$

$\therefore \tfrac{17}{20}$ **is the bigger amount and is the one Mary should choose.** (Note that the *size* of the prize does not matter here.)

Simplify:

1. $\frac{3}{4}+\frac{1}{2}$ **2.** $\frac{2}{3}-\frac{1}{2}$ **3.** $3\frac{1}{2}+\frac{3}{5}$ **4.** $2\frac{1}{2}-1\frac{3}{8}$ **5.** $2\frac{1}{4}-\frac{1}{2}$

6. $3\frac{1}{3}+2\frac{3}{4}$ **7.** $4\frac{3}{5}-2\frac{3}{4}$ **8.** $\frac{51}{14}-\frac{29}{21}$ **9.** $2\frac{1}{3}+1\frac{2}{5}-3\frac{1}{4}$

10. $\frac{3}{10}-\frac{7}{20}+1\frac{1}{100}$ **11.** $1+\frac{1}{2}+\frac{1}{4}+\frac{1}{8}+\frac{1}{16}+\frac{1}{32}$ **12.** $7\frac{1}{3}-3\frac{5}{12}-2\frac{3}{4}$.

5. Multiplication of Fractions

When two or more fractions are multiplied together, we may cancel out factors which appear one in the numerator and one in the denominator.

$$\text{For example} \quad \frac{2}{3}\times\frac{3}{4} = \frac{\overset{1}{2}}{3}\times\frac{\overset{1}{3}}{\underset{2}{4}} = \frac{1}{2}$$

because the 3's are common, and 2 is also a common factor.

To understand this properly it is necessary to consider the process in two stages:

(i) $\frac{1}{3}\times\frac{3}{4} = \frac{1}{3}$ of $\frac{3}{4} = \frac{3}{12} = \frac{1}{4}$;

(ii) $\frac{2}{3}\times\frac{3}{4} = 2$ of $\left(\frac{1}{3}\times\frac{3}{4}\right) = 2\times\frac{1}{4}$ (from (i))
$= \frac{1}{4}+\frac{1}{4} = \frac{2}{4} = \frac{1}{2}$.

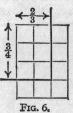

FIG. 6.

Fig. 6 shows 12 squares, laid out 3 by 4. If we take $\frac{2}{3}$ of the columns $\times\frac{3}{4}$ of the rows we get 6 squares, i.e. $\frac{1}{2}$ of the area.

When numerators and denominators do not cancel, we may simplify as follows:

$$\frac{2}{7}\times\frac{3}{5} = \frac{2\times3}{7\times5} = \frac{6}{35}.$$

We may also use both ideas, as in the quantity

$$\frac{7}{9}\times\frac{3}{4} = \frac{7}{\underset{3}{9}}\times\frac{\overset{1}{3}}{4} = \frac{7}{3\times4} = \frac{7}{12}.$$

Ex. 11. Simplify $\frac{7}{10}\times\frac{5}{21}$.

$$\frac{7}{10}\times\frac{5}{21} = \frac{\overset{1}{7}\times\overset{1}{5}}{\underset{2}{10}\times\underset{3}{21}} = \frac{1}{6}.$$

When whole number parts appear in *multiplication* of fractions (i.e. mixed numbers occur) the *mixed numbers* must first be converted to *improper fractions* before simplification can take place. This also applies to *division*.

Ex. 12. Simplify $2\frac{1}{4} \times 5\frac{1}{3}$.

$$2\frac{1}{4} \times 5\frac{1}{3} = \frac{\overset{3}{\cancel{9}}}{\underset{1}{\cancel{4}}} \times \frac{\overset{4}{\cancel{16}}}{\underset{1}{\cancel{3}}} = \frac{12}{1} = \mathbf{12}.$$

$$(2\frac{1}{4} = \frac{8}{4} + \frac{1}{4} = \frac{9}{4}; \quad 5\frac{1}{3} = \frac{15}{3} + \frac{1}{3} = \frac{16}{3})$$

Ex. 13. Simplify $1\frac{1}{2} \times 2\frac{2}{3} \times \frac{7}{16}$.

$$1\frac{1}{2} \times 2\frac{2}{3} \times \frac{7}{16} = \frac{3}{2} \times \frac{\overset{1}{\cancel{8}}}{\underset{1}{\cancel{3}}} \times \frac{7}{\underset{2}{\cancel{16}}} = \frac{7}{4} = \mathbf{1\frac{3}{4}}.$$

Ex. 14. Find the value of $\left(\frac{2}{3}\right)^3 \times 67\frac{1}{2}p$.

We have already seen that the index 3 to a number means that it occurs 3 times.

$$\therefore \left(\frac{2}{3}\right)^3 \times 67\frac{1}{2}p = \frac{2}{3} \times \frac{2}{3} \times \frac{2}{3} \times \frac{135}{2}p$$

$$= \frac{\overset{4}{\cancel{8}}}{\underset{3}{\cancel{27}}} \times \frac{\overset{5}{\overset{15}{\cancel{135}}}}{\underset{1}{\cancel{2}}}p$$

$$= \mathbf{20p.}$$

6. Division of Fractions

Consider $\frac{3}{7} \div \frac{2}{5}$. It is not obvious what this means. We could, however, write the quantity as follows:

$$\frac{3}{7} \div \frac{2}{5} = \frac{\frac{3}{7}}{\frac{2}{5}} \begin{array}{l} \leftarrow \text{numerator} \\ \leftarrow \text{denominator} \end{array}$$

Now if we multiply top and bottom by the same amount we do not change the quantity.

$$\therefore \frac{\frac{3}{7}}{\frac{2}{5}} = \frac{\frac{3}{7} \times 5 \times 7}{\frac{2}{5} \times 5 \times 7} = \frac{3 \times 5}{2 \times 7} = \frac{3 \times 5}{7 \times 2} = \frac{3}{7} \times \frac{5}{2}.$$

We know how to work out the final result (from paragraph 5 above), but the interest lies in the facts that we have *inverted the second expression and multiplied by it* in our steps of reasoning. This leads to the following rule.

When dividing by a fraction, invert it and multiply by the result,

e.g. $\frac{2}{5} \div \frac{9}{11} = \frac{2}{5} \times \frac{11}{9}$; $\frac{4}{7} \div 3 = \frac{4}{7} \times \frac{1}{3}$; $\frac{6}{17} \div \frac{1}{4} = \frac{6}{17} \times 4$.

Note that dividing by 3 is the same as multiplying by $\frac{1}{3}$, for the *reciprocal* (i.e. inverted form of $\frac{3}{1}$ is $\frac{1}{3}$). Similarly the reciprocal of $\frac{1}{4}$ is $\frac{4}{1}$, i.e. 4.

Ex. 15. Simplify $\frac{4}{15} \div \frac{3}{5}$.

$$\frac{4}{15} \div \frac{3}{5} = \frac{4}{\overset{}{\underset{3}{15}}} \times \frac{\overset{1}{5}}{3} = \frac{4}{9}.$$

Ex. 16. Simplify $2\frac{1}{7} \div 1\frac{1}{4}$.

$$2\frac{1}{7} \div 1\frac{1}{4} = \frac{15}{7} \div \frac{5}{4} = \frac{\overset{3}{15}}{7} \times \frac{4}{\underset{1}{5}} = \frac{12}{7} = 1\frac{5}{7}.$$

Ex. 17. Find the value of $2\frac{1}{2} \times 3\frac{2}{3} \div 1\frac{4}{7}$.

$$2\frac{1}{2} \times 3\frac{2}{3} \div 1\frac{4}{7} = \frac{5}{2} \times \frac{11}{3} \div \frac{11}{7} = \frac{5}{2} \times \frac{\overset{1}{11}}{3} \times \frac{7}{\underset{1}{11}} = \frac{35}{6} = 5\frac{5}{6}.$$

EXERCISE 3

Simplify:

1. $\frac{4}{7} \times \frac{2}{3}$ 2. $\frac{3}{8} \times 1\frac{5}{9}$ 3. $1\frac{2}{5} \times 2\frac{2}{7}$ 4. $(\frac{3}{4})^2$
5. $2\frac{1}{5} \times 3\frac{3}{4} \times 4\frac{1}{8}$ 6. $(\frac{2}{5})^2 \times 12\frac{1}{2}$ 7. $(\frac{2}{3})^3 \times (\frac{3}{4})^2$ 8. $\frac{3}{4} \div \frac{2}{3}$
9. $1 \div \frac{3}{8}$ 10. $2 \div 1\frac{1}{3}$ 11. $\frac{3}{4} \div 1\frac{1}{2}$ 12. $2\frac{3}{7} \div 2\frac{1}{10}$
13. $3 \times \frac{2}{5} \div 3\frac{1}{3}$ 14. $1\frac{2}{5} \times 3\frac{3}{7} \div 2\frac{1}{2}$.

7. Mixed Signs

In simplifying fractions in which brackets and other signs occur we merely apply the mnemonic BODMAS given in Chapter 2.

Ex. 18. Evaluate $\frac{1}{2} \div \frac{1}{3} - 1\frac{1}{4}$.

$\frac{1}{2} \div \frac{1}{3} - 1\frac{1}{4} = \frac{1}{2} \times 3 - 1\frac{1}{4}$ (*Divide*, before *Subtracting*)

$= \frac{3}{2} - 1\frac{1}{4} = 1\frac{1}{2} - 1\frac{1}{4}$

$= \frac{1}{2} - \frac{1}{4} = \frac{2-1}{4} = \frac{1}{4}.$

Care is needed in applying the rules to these mixed cases and, although short cuts are often possible, they are not advised until experience is gained.

Ex. 19. Simplify $2\frac{1}{2} \times 1\frac{1}{3} - \frac{3}{5} \div 1\frac{4}{11}$.

$2\frac{1}{2} \times 1\frac{1}{3} - \frac{3}{5} \div 1\frac{4}{11} = \frac{5}{2} \times \frac{4}{3} - \frac{3}{5} \div \frac{15}{11}$ (Improper Fractions)

$$= \frac{5}{2} \times \overset{2}{\underset{}{\frac{4}{3}}} - \frac{3}{5} \times \frac{11}{\underset{15}{\cancel{15}}} \quad \text{(Division)}$$

$$= \overset{1}{\underset{}{\frac{10}{3}}} - \overset{5}{\underset{}{\frac{11}{25}}} \quad \text{(Multiplication)}$$

$$= 3\frac{1}{3} - \frac{11}{25} = 3 + \frac{25 - 33}{75} \quad \text{(Subtraction)}$$

$$= 2 + \frac{75 + 25 - 33}{75} = 2 + \frac{67}{75}$$

$$= 2\frac{67}{75}.$$

Ex. 20. Simplify $1\frac{1}{2} \div (2\frac{4}{9} \times 1\frac{4}{11})$.

Here the bracket has complete priority.

$$2\frac{4}{9} \times 1\frac{4}{11} = \frac{\overset{2}{\cancel{22}}}{9} \times \frac{\overset{5}{\cancel{15}}}{11} = \frac{10}{3}$$

$$\therefore \ 1\frac{1}{2} \div (2\frac{4}{9} \times 1\frac{4}{11}) = 1\frac{1}{2} \div \frac{10}{3} = \frac{3}{2} \div \frac{10}{3}$$

$$= \frac{3}{2} \times \frac{3}{10} = \frac{9}{20}.$$

Ex. 21. Simplify $\dfrac{3\frac{1}{3} + 1\frac{1}{4} \text{ of } \frac{2}{5}}{2\frac{2}{5} - 1\frac{1}{4}}$.

First simplify numerator and denominator as far as possible.

Numerator $= 3\frac{1}{3} + \frac{5}{4} \times \frac{2}{5} = 3\frac{1}{3} + \frac{1}{2} = 3\frac{5}{6}$.

Denominator $= 2\frac{2}{5} - 1\frac{1}{4} = 1 + \frac{2}{5} - \frac{1}{4} = \dfrac{20 + 8 - 5}{20}$

$$= \frac{23}{20}.$$

$\therefore$ Expression $= 3\frac{5}{6} \div \frac{23}{20} = \frac{\overset{1}{\cancel{23}}}{6} \times \frac{\overset{10}{\cancel{20}}}{\underset{1}{\cancel{23}}} = \frac{10}{3}$

$$= 3\frac{1}{3}.$$

EXERCISE 4

Simplify the following expressions:

1. $\frac{2}{3} + \frac{3}{7} \times 1\frac{2}{5}$ **2.** $\dfrac{\frac{2}{3} + \frac{3}{4}}{\frac{3}{4} - \frac{2}{3}}$ **3.** $(2\frac{1}{2} - 1\frac{3}{5}) \div (3\frac{1}{4} + 1\frac{1}{7})$

4. $\dfrac{\frac{3}{5} \times 1\frac{1}{4} - \frac{2}{5}}{1\frac{1}{2} \times 2\frac{1}{3} + 7\frac{7}{8}}$ **5.** $\frac{7}{8} \times (3\frac{1}{4} - 2\frac{2}{7}) + \frac{1}{2}$ **6.** $\dfrac{\frac{1}{2} + \frac{2}{5} - \frac{3}{8}}{\frac{1}{2} + \frac{2}{5} + \frac{3}{8}}$

8. Problems

We shall complete the chapter with a few applications to problems.

Ex. 22. Williams, Brown and Higginbottom share a prize of £120. Williams is entitled to one-third of it and Brown gets three-eighths of it. How much money can Higginbottom expect to receive?

Brown and Williams have $\frac{3}{8} + \frac{1}{3}$ of the money

∴ Higginbottom will have $1 - (\frac{3}{8} + \frac{1}{3})$ of it.

$$1 - (\tfrac{3}{8} + \tfrac{1}{3}) = 1 - \frac{9 + 8}{24} = 1 - \tfrac{17}{24}$$

$$= \frac{24 - 17}{24} = \tfrac{7}{24}$$

∴ Higginbottom gets $\frac{7}{24}$ of £120

$$= £\frac{7}{24} \times \overset{\overset{5}{\cancel{10}}}{\cancel{120}} = \textbf{£35.}$$

Ex. 23. Roberts died and left two-thirds of his estate to his wife; three-quarters of the remainder was willed to his son. If his niece was entitled to the residue, how much did she receive if the estate was valued at £4800?

The wife received $\frac{2}{3}$ of the estate, so $\frac{1}{3}$ remained.

The son received $\frac{3}{4}$ of this remainder, i.e. $\frac{3}{4}$ of $\frac{1}{3} = \frac{1}{4}$ of the estate.

Between them they had $\frac{2}{3} + \frac{1}{4} = \frac{11}{12}$ of the estate.

∴ The niece received what was left, i.e. $\frac{1}{12}$.

∴ The niece had $\frac{1}{12}$ of £4800 = **£400.**

EXERCISE 5

1. Express $46\frac{1}{2}$p as a fraction of $77\frac{1}{2}$p in lowest terms.

2. Two-fifths of a sum on money is £2400. How much is three-quarters of it?

3. Brown sold his car for three-fifths of the cost price and finds that in doing so he has lost £750. For how much did he sell the car?

4. The formula for the focal length of a lens is $\dfrac{1}{v} + \dfrac{1}{u} = \dfrac{1}{f}$. Find f if

$u = 27$ cm, $v = 30$ cm.

5. A contractor employs 119 men to build a factory. After 44 days two-thirds of the building is completed. He then employs 35 more men. How long will it take to complete the work?

6. Robinson cycles 2 kilometres to the station. He takes a train for.two-thirds of his journey and a bus for the remaining quarter. How long was his complete journey?

7. Brown can plant his lettuces in an hour and a half. His son could plant them in two and a half hours. How long would it take them if they worked together?

8. Find the cost of $1\frac{1}{4}$ kg ham at £2·34 per kilogramme.

CHAPTER FIVE

DECIMALS

1. Introduction

We have seen that fractions consist of quantities like $\frac{3}{4}$, $\frac{5}{19}$ and $\frac{7}{2}$, and that they are cumbersome to handle, because simplification entails bringing them to the same denominator or some other fairly elaborate manipulation. A simpler process, involving uniform denominators, is clearly often desirable.

The system of *numbers* we use is the decimal scale (i.e. scale of 10). The numbers read 1, 10, 100, 1000 and so on. This suggests an extension

$$\leftarrow 1000,\ 100,\ 10,\ 1,\ \frac{1}{10},\ \frac{1}{100},\ \frac{1}{1000} \rightarrow$$

so that the terms on the left of 1 are obtained by successive multiplication by ten, and those on the right by successive division by ten. We can, if we wish, rewrite the sequence above as

$$\leftarrow 10^3,\ 10^2,\ 10,\ 1,\ \frac{1}{10}\ \frac{1}{10^2},\ \frac{1}{10^3} \rightarrow$$

that is, as *powers* of ten. The fractions on the right all have powers of ten as denominators. They are called *decimals*, so a decimal is defined as a *fraction whose denominator is some power of ten.*

We greatly simplify working with decimals by adopting the following notation:

$$(1\ \text{tenth})\ \tfrac{1}{10} = 0\cdot 1; \quad (1\ \text{hundredth})\ \tfrac{1}{100} = 0\cdot 01;$$
$$(3\ \text{hundredths})\ \tfrac{3}{100} = 0\cdot 03;$$
$$(17\ \text{thousandths})\ \tfrac{17}{1000} = 0\cdot 017.$$

Ex. 1. Express the following as decimals:

(a) $4 + \frac{8}{100}$ (b) $20 + 8 + \frac{7}{10}$

(c) $3 \times 10^4 + 2 \times 10^2 + 4 \times 10$ (d) $\frac{6}{10} + \frac{5}{100}$

(e) $74\,905 + \frac{87}{1000}$ (f) $\frac{3}{1000}$.

The results have been laid out in tabular form so as to illustrate clearly the various points to bear in mind. It is quite unnecessary to adopt this method in working examples once the principles are understood.

	10^4	10^3	10^2	10	1	$\frac{1}{10}$	$\frac{1}{10^2}$	$\frac{1}{10^3}$	
	Ten-thousands	Thousands	Hundreds	Tens	Units	Tenths	Hundredths	Thousandths	Answers
(a)					4		8		4·08
(b)				2	8	7			28·7
(c)	3		2	4					30 240
(d)						6	5		0·65
(e)	7	4	9		5		8	7	74 905·087
(f)								3	0·003

Notes

1. We put in noughts to fill gaps in the results.

2. In (c) above, the number was written in powers of ten to show that it can be dealt with as a decimal.

3. Note that in (e) $\frac{87}{1000}$ is the same as

$$\frac{80}{1000} + \frac{7}{1000} = \frac{1}{100} + \frac{7}{1000}.$$

4. In (d) and (f) where there is no whole number in front of the decimal point, we put in a 0.

5. Figures in the tenths, hundredths, thousandths etc. column are called figures in the first, second, third etc. decimal place.

2. Addition and Subtraction

We use the decimal point as our guide in putting down numbers to be added or subtracted in columns.

Ex. 2. Add together 72·95, 2·0604, 319·5, 204.

$$
\begin{aligned}
72 &\cdot 95 \\
2 &\cdot 0604 \\
319 &\cdot 5 \\
204 & \\
\hline
\mathbf{598} &\cdot \mathbf{5104}
\end{aligned}
$$

The carrying figures are taken from column to column in the ordinary way (as in Chapter 1). The presence of the decimal point does not affect the procedure.

Ex. 3. Subtract 4·0805 from 9·1.

$$9·1000$$
$$4·0805$$
$$\overline{5·0195}$$

Here, unlike addition, it is best to put in the missing zeros (9·1 is the same as 9·1000).

1. Express the following as decimals:
   (a) $2+\frac{3}{10}$ (b) $6+\frac{1}{100}$ (c) $30+7+\frac{5}{10}+\frac{7}{100}$ (d) $\frac{9}{10}+\frac{1}{100}$
   (e) $\frac{7}{100}$ (f) $\frac{1}{10000}$ (g) $80+5+\frac{7}{100}+\frac{6}{1000}$.

2. Write out the following decimals as the sum of decimal fractions (e.g. $2·04 = 2+\frac{4}{100}$):
   (a) 6·8   (b) 14·92   (c) 207·04
   (d) 0·6   (e) 0·094   (f) 1000·001.

3. Add together:
   (a) 26·48, 3·9, 0·882   (b) 300·7, 0·006, 29·908
   (c) 217·85, 86·57, 1709·08   (d) 2·0064, 693·907, 0·087, 774·8.

4. Subtract:
   (a) 38·7 from 62·4   (b) 3·95 from 17·7
   (c) 0·927 from 10   (d) 904·767 from 1202·6825.

3. Multiplication and Division—Introduction

Multiplying by 10 merely necessitates moving the decimal point *one place to the right*.

E.g. $4·37 \times 10 = 43·7$ (because $[4+\frac{3}{10}+\frac{7}{100}] \times 10$
$$= 40+3+\frac{7}{10} = 43·7).$$

Dividing by 10 is carried out by moving the decimal point *one place to the left*.

E.g. $9·2 \div 10 = 0·92$ (because $[9+\frac{2}{10}] \div 10 = \frac{9}{10}+\frac{2}{100} = 0·92$).

This rule can be generalised as follows:

To multiply by 10, 100, 1000 etc., move the decimal point 1, 2, 3 etc. places to the right.

To divide by 10, 100, 1000 etc., move the decimal point 1, 2, 3 etc. places to the left.

$$E.g. \ 62·5 \times 1000 = 62\ 500$$
$$62·5 \div 1000 = 0·0625.$$

Multiplication and division in general are carried out in much the same way as with ordinary numbers in Chapter 1. Care must be taken to see that the decimal point is kept in the correct position.

This is easily effected by noting the total number of decimal places present and putting the decimal point in front of them.

Consider 0.035×3.

We have $35 \times 3 = 105$

i.e. $0.035 \times 3 = \mathbf{0.105}$ (as there are 3 decimal places).

We can check this as follows:

$$0.035 \times 3 = \tfrac{35}{1000} \times 3 = \tfrac{105}{1000} = \mathbf{0.105}.$$

Ex. 4. $0.7 \times 3 = \mathbf{2.1} \, [0.7 \times 3 = \tfrac{7}{10} \times 3 = \tfrac{21}{10} = 2.1].$

Ex. 5. $21.6 \times 7 = \mathbf{151.2}.$

Ex. 6. $0.0984 \times 6 = \mathbf{0.5904}.$

Ex. 7. $0.0042 \times 5 = 0.0210 = \mathbf{0.021}.$

Although we get 0.0210, *eventually* the final zero is not put down as it does not affect the value of the answer.

4. Multiplication

When decimals appear in both the multiplicand and the multiplier:

(a) *Multiply the numbers as though they were whole.*

(b) *Determine the position of the decimal point in the answer by finding the total number of decimal places.*

Ex. 8. $0.073 \times 0.1 = \mathbf{0.0073}$
$(3+1 = 4$ decimal places$)$.

Ex. 9. Find the value of 23.58×17.5.

Use the rules above.

```
  2 358
    175 ×
  ───────
  11 790     2+1 = 3 decimal places.
  165 06     Result 412·650; we then omit the final zero when
  235 8      writing the answer, but not before.
  ───────
  412 650
```

$\therefore \ 23.58 \times 17.5 = \mathbf{412.65}.$

Although the above is undoubtedly the simplest method and also the safest, it is not suitable when contracted multiplication is required (i.e. when several of the figures in the answer are superfluous).

Multiplication by factors can often be carried out, but it is rarely of value. (Division by factors is, on the other hand, important.)

As a final example of multiplication of the normal type consider:

Ex. 10.   36 000 × 0·018 35

= 36 × 18·35

= **660·6.**

$$\begin{array}{r} 1\ 835 \\ 36 \times \\ \hline 11\ 010 \\ 55\ 05 \\ \hline \mathbf{66\ 060} \end{array}$$

[We divide 36 000 by 1000, thus removing the zeros; we multiply 0·018 35 by 1000, so compensating. The result is considerable simplification.

Otherwise we can proceed as follows:

36 00 × 1835 = 1 835 00 × 36 = 66 060 000
5 d.p. leave 660·600 00 = **660·6**, but this is clumsy.

We always multiply by the number which has fewer figures in it; i.e. by 36, not by 1835.]

5. Division

There are numerous methods of division in use, but, for the student practising alone, the following are sufficient.

Ex. 11. Divide 21·7 by 0·0014.

$$\frac{21·7}{·0014} = \frac{217\ 000}{14}$$

= **15 500.**

$$\begin{array}{r} 7)217\ 000 \\ \hline 2)\ 31\ 000 \\ \hline \mathbf{15\ 500} \end{array}$$

We *multiply* the top and bottom by the same power of 10 (in this case, we multiply by 10 000) to make the divisor a whole number; i.e. we move the decimal point *the same number of places to the right* in the numerator and denominator.

Ex. 12. Divide 380·64 by 2400.

$$\frac{380·64}{2400} = \frac{3·8064}{24}$$

= **0·1586.**

$$\begin{array}{r} 6)3·8064 \\ \hline 4)0·6344 \\ \hline \mathbf{0·1586} \end{array}$$

Here we *divide* the top and bottom by the same power of 10 to simplify the working; i.e. we move the decimal point *the same number of places to the left* in the numerator and denominator.

It often happens that an exact answer cannot be obtained, or, even if it can be obtained, that it contains more decimal places than we need for our purposes. In such a case we approximate our ans r to the degree of accuracy we require. Consider the following example:

Ex. 13. Find the value of $0 \cdot 0867 \div 0 \cdot 21$.

$$\frac{0 \cdot 0867}{0 \cdot 21} = \frac{8 \cdot 67}{21}$$

$$= 0 \cdot 412 \ldots$$

$$\backsimeq \mathbf{0 \cdot 41}.$$

$$7)\overline{8 \cdot 670}$$
$$3)\overline{1 \cdot 238 \ldots}$$
$$\mathbf{0 \cdot 412} \ldots$$

(The symbol $\backsimeq$ means 'is approximately equal to'.)

What we have done is to work out the question to 3 places of decimals and we then corrected the answer to an accuracy of 2 decimal places (2 d.p.).

If we had wanted the answer correct to 3 d.p. we would have worked a little farther:

$$7)\overline{8 \cdot 6700}$$
$$3)\overline{1 \cdot 2385 \ldots}$$
$$\mathbf{0 \cdot 4128} \ldots$$

and we would have found that $\dfrac{0 \cdot 0867}{0 \cdot 21} \backsimeq \mathbf{0 \cdot 413}$ (3 d.p.).

We take the value $0 \cdot 413$ because $0 \cdot 4128$ lies between $0 \cdot 412$ and $0 \cdot 413$, and *is nearer to the latter*.

One case merits particular attention. What do we take if the last figure is a 5 and we wish to approximate?

E.g. $0 \cdot 6025$ might be $0 \cdot 602$ or $0 \cdot 603$ (to 3 d.p.).

We merely take a convention to go to the higher number, i.e.:

$$0 \cdot 6025 \backsimeq 0 \cdot 603 \text{ (correct to 3 decimal places).}$$

In all of the examples 11–13, we were able to divide by factors. We now have to study long division. As before, we shall make the divisor a whole number.

Ex. 14. Divide $3 \cdot 8$ by $0 \cdot 89$ and give the answer correct to 2 d.p.

$$\frac{3 \cdot 8}{0 \cdot 89} = \frac{380}{89}$$

$$= 4 \cdot 269 \ldots$$

$$\backsimeq \mathbf{4 \cdot 27} \text{ (2 d.p.).}$$

$$\begin{array}{r} \mathbf{4 \cdot 269} \\ 89)\overline{380 \cdot 000} \\ \underline{356} \\ 24 \cdot 0 \\ \underline{17 \cdot 8} \\ 6 \cdot 20 \\ \underline{5 \cdot 34} \\ 860 \\ \underline{801} \end{array}$$

Notice that the quotient is written over the dividend so that the quotient decimal point comes over the dividend decimal point.

It will be observed that in Examples 13 and 14 we introduce zeros as necessary in the dividend. They do not affect its value.

In Ex. 14 we write 380 as 380·000, introducing the decimal point and as many zeros as we need in the working. $380 \div 89 = 4$, remainder 24. Bring down the first decimal place figure, i.e. 0, *and enter the decimal point in the quotient.* $240 \div 89 = 2$, remainder 62, etc. If we write the quotient over the dividend we see clearly where the decimal point will lie.

In the above work we have approximated to a specified number of *places of decimals*. We sometimes require an answer correct to a particular number of *significant figures*. To understand the meaning of significant figures (or significant digits, as they may more accurately be called) requires care.

Digits are the numbers 0, 1, 2, 3, . . ., 9, i.e. the positive whole numbers less than 10, together with zero. Significant digits, or significant figures, are the digits of a number beginning with the first digit on the left which is not zero and ending with the last digit on the right which is not zero. Some illustrations will help the reader to understand this.

$$
\begin{array}{llll}
3608 & \text{has } 4 \text{ significant figures} \\
3600 & \text{,, } 2 & \text{,,} & \text{,,} \\
0·045 & \text{,, } 2 & \text{,,} & \text{,,} \\
10\ 800 & \text{,, } 3 & \text{,,} & \text{,,} \\
0·1453 & \text{,, } 4 & \text{,,} & \text{,,}
\end{array}
$$

The value of 0·4277 correct to 2 significant figures (2 s.f.) is 0·43.

Likewise $4315 \simeq 4300$ (2 s.f.)
$4360 \simeq 4400$ (2 s.f.).

Note that there are cases where end zeros may be significant:

$$21·769\ 63 \simeq 21·770 \text{ (5 s.f.)}$$
(for 0·769 63 is nearer 0·770 than 0·769).

EXERCISE 2

Write down the following numbers correct to the accuracy required:

1. 417.3 (3 s.f.) **2.** 0.0868 (3 d.p.) **3.** 22·0097 (3 d.p.)
4. 6·089 (2 s.f.) **5.** 0·006 993 (4 d.p.) **6.** 0·030 87 (3 s.f.)
7. 9·843 05 (3 s.f.) **8.** 0·0984 (3 d.p.).

Ex. 15. Calculate $0·324 \div 72·5$ to 3 significant figures.

$$\frac{0·324}{72·5} = \frac{3·24}{725}$$

$$= 0·004\ 468 \ldots$$

$$= \mathbf{0·004\ 47}\ (3\ \text{s.f.}).$$

```
           0·004 468
     725)3·240 000
         2·900
         ────
           340 0
           290 0
           ─────
            50 00
            43 50
            ─────
             6 500
             5 800
```

725 will not divide into 3; put in decimal (0·). It will not divide into 3·2; put in 0·0. It will not divide into 3·24; put in 0·00. It *will* divide into 3·240; enter first significant figure, 4.

EXERCISE 3

1. Find the value of:

(a) $0·04 \times 10$	(b) $0·63 \times 100$	(c) $3·85 \times 1000$	(d) $0·007 \times 1000$
(e) $0·05 \times 3$	(f) $0·59 \times 12$	(g) $0·08 \times 20$	(h) $1·64 \times 300$
(i) $0·042 \times 120$	(j) $28·9 \times 70$	(k) $2·024 \times 110$	(l) $0·53 \times \frac{1}{100}$.

2. Find the value of:

(a) $6·8 \div 2$	(b) $5·8 \div 4$	(c) $3·25 \div 10$	(d) $0·08 \div 100$
(e) $3·84 \div 20$	(f) $79·8 \div 70$	(g) $0·0965 \div 0·5$	(h) $0·825 \div 0·15$
(i) $297 \div 0·055$	(j) $1·62 \div 0·009$	(k) $0·146\ 08 \div 40$	(l) $0·283 \div \frac{1}{100}$.

3. Find the value of the following exactly, unless otherwise stated:

(a) $24·7 \times 3·8$	(b) $6·8 \times 0·042$	(c) $386·2 \times 0·072$
(d) $8500 \times 0·001\ 32$	(e) $79·7 \times 31·4$ (2 s.f.)	(f) $8304 \times 0·061$ (3 s.f.)
(g) $8200 \times 0·439$	(h) $9070 \times 0·001\ 94$ (4 s.f.)	(i) $0·23 \times 61·86$ (2 d.p.).

4. Find the value of the following, correct to the accuracy named:

(a) $29·4 \div 33$ (2 d.p.)	(b) $3·046 \div 42$ (3 d.p.)
(c) $0·987 \div 0·71$ (3 s.f.)	(d) $20·45 \div 0·93$ (3 s.f.)
(e) $6·4 \div 760$ (4 d.p.)	(f) $39·84 \div 1·773$ (3 d.p.).

6. Fractions and Decimals

Some fractions can be expressed as exact decimals.

$\frac{1}{2} = 0·5,\ \frac{1}{4} = 0·25,\ \frac{1}{8} = 0·125.$

```
4)1·00
  0·25
```

$\frac{3}{4} = 3 \times \frac{1}{4} = 3 \times 0·25 = 0·75$ and so on.

Others, however, cannot.

$\frac{1}{3} = 0·333 \ldots$

```
3)1·000
  0·333 ...
```

Ex. 16. Express $\frac{3}{7}$ as a decimal to 3 places.

$\frac{3}{7} \simeq 0.429$ $7\overline{)3.000\ 00}$

                      $0.428\ 57\ldots$

The reverse process is easily mastered.

Ex. 17. Express 0.16 as a fraction in lowest terms.

$$0.16 = \frac{16}{100} = \frac{\overset{4}{\cancel{16}}}{\underset{25}{\cancel{100}}} = \frac{4}{25}.$$

EXERCISE 4

1. Express the following fractions as decimals, exactly where possible, but to four decimal places otherwise:

   (a) $\frac{3}{5}$ (b) $\frac{5}{8}$ (c) $\frac{7}{11}$ (d) $\frac{2}{9}$ (e) $\frac{7}{18}$ (f) $\frac{9}{17}$.

2. Convert the following decimals to vulgar fractions in lowest terms:

   (a) 0.12 (b) 0.716 (c) 0.085 (d) 0.0906 (e) 7.1625.

3. Express the following fractions as decimals correct to three places and hence arrange them in ascending order of magnitude: $\frac{171}{50}$, $\frac{147}{43}$, $\frac{427}{125}$.

4. Find the value of: (a) $\dfrac{2.5 \times 0.64}{40}$ (b) $\dfrac{0.063 \times 7.8}{0.0091}$.

5. Find the value of:

   (a) $(2.54)^2$ (b) $(0.02)^2$ (c) $(0.11)^3$.

6. Find, correct to 2 significant figures, the value of:

   (a) $(\frac{3}{7})^2$ (b) $(0.343)^3 \div (0.063)^2$.

CHAPTER SIX

THE METRIC SYSTEM

1. Origin of the Metric System

Preliminary work on the metric system of measurement began in France as early as 1789. The aim was to simplify the complicated collection of weights and measures, which were in general use, by adopting a single *standard measure* for each of length, area, volume and weight (which should, incidentally, be more accurately referred to as *mass*).

The system came into use as a result of a recommendation in 1791, by a committee of the French Academy, that the *metre* (after which the whole structure of measurement is named) be adopted as the standard measure of length. The metre was originally defined as one ten-millionth (i.e. 10^{-7}) part of the earth's meridional quadrant through Paris. A later measurement was that of the length at 0°C between the centres of two lines engraved on a special platinum-iridium bar in Paris. In 1960, a complex scientific definition ensued.

It will be observed that the introduction of the metric system coincided with the French Revolution. Such an occasion would be ripe for the establishment of a revolutionary idea having beneficial intent. By 1840 the system was obligatory in France, except with regard to time measurement, wherein the sole modification which has come to pass is the gradual introduction of the 24-hour clock throughout the civilized world. Time still remains sexagesimal in concept (see page 2).

2. The Metric System and Standard International (SI) Units

The United Kingdom is fully committed to utilising the metric system, in common with Europe and much of the rest of the world. It is clearly desirable that, at the same time, there shall be full standardisation of all units of measurement among nations. This has led to the creation of Standard International Units[5] (or, more briefly, SI units), the simpler ones of which are referred to in this book. Foreign trade and scientific work, the second of which recognises few state frontiers, will be greatly facilitated.

We shall first look at the metric system and its prefixes, which determine the size of a measurement (in multiples of 10). Indication

[5] The agreed system is really called *le Système International d'Unités*.

will then be given as to *which prefixes* have gained favour in the system of SI units.

3. Length

The unit of length, as already stated in Section 1 above, is the *metre* (m). It is rather more than a yard, being approximately 39.37 in. For some purposes it can be considered as nearly 1.1 yd, the error being about 1 in 2000, for

$$1.1 \text{ yd} = 36 \times 1.1 \text{ in} = 39.6 \text{ in} \frown 1 \text{ m}.$$

Other lengths are obtained by inserting prefixes, based on Greek words for larger amounts than the *standard* unit and on Latin for smaller amounts, e.g.

$$1000 \text{ metres} = 1 \; kilo\text{metre (km)}$$
$$100 \text{ metres} = 1 \; hecto\text{metre (hm)}$$
$$10 \text{ metres} = 1 \; deka\text{metre (Dm)}$$
$$0.1 \text{ metre} = 1 \; deci\text{metre (dm)}$$
$$0.01 \text{ metre} = 1 \; centi\text{metre (cm)}$$
$$0.001 \text{ metre} = 1 \; milli\text{metre (mm)}$$

The same *prefixes* are used for the measures of area, volume and weight.

Only in the case of the *deka* prefix abbreviation is a capital letter needed (to save confusion with *deci*, which is much smaller).

It will be noticed that the prefixes proceed in multiples of 10, this being intended to link up directly with decimals.

$$10 \text{ mm} = 1 \text{ cm}; \; 10 \text{ cm} = 1 \text{ dm}; \; 10 \text{ dm} = 1 \text{ m};$$
$$10 \text{ m} = 1 \text{ Dm}; \; 10 \text{ Dm} = 1 \text{ hm}; \; 10 \text{ hm} = 1 \text{ km}.$$

It is worth digressing a moment on to the method of multiplying powers of 10 and thereby anticipating Ch. 12, pp. 118–133, on logarithms and indices. We have, for example,

$$10^3 \times 10^2 = 1000 \times 100 = 100\ 000 = 10^5,$$

the result showing that we could have added the powers of 10, for $2+3 = 5$.

Similarly $\qquad\qquad 10^3 \times \frac{1}{10} = 10^2$

suggesting that we replace $\frac{1}{10}$ by 10^{-1}, which would give $3-1 = 2$, leading to

$$10^3 \times 10^{-1} = 10^2.$$

In fact this always works (provided that we *define* $10^0 = 1$). The process is later explained in greater detail.

Examples

$1 \text{ km} = 10^3 \text{ m} = 10^3 \times 10^2 \text{ cm} = 100\ 000 \text{ cm};$
$1 \text{ km} = 10^3 \text{ m} = 10^3 \times 10^3 \text{ mm} = 1\ 000\ 000 \text{ mm}.$

In accordance with SI Units, length will be measured in the *millimetre* (mm), *metre* (m) $\triangleq$ 39·37 in, *kilometre* (km) $\triangleq$ 0·6214 mile; also, because of its convenient size for everyday life, *centimetre* (cm) $\triangleq$ 0·3937 in. These are the only ones which need to be memorised for general purposes, although the others may be met from time to time.

SI Unit	Length in metres	Approximate Length in Imperial units
1 kilometre	$10^3 \text{ m} = 1000 \text{ m}$	0·6214 mile
1 metre	1 m	39·37 in
1 centimetre	$10^{-2} \text{ m} = \frac{1}{100} \text{ m}$	0·3937 in
1 millimetre	$10^{-3} \text{ m} = \frac{1}{1000} \text{ m}$	0·0394 in

Ex. 1. Taking $1 \text{ m} \triangleq 39 \cdot 37$ in, show that 1 km is very nearly $\frac{5}{8}$ mile. Find the error involved, giving the answer in feet, correct to the nearest foot.

We have

$$1 \text{ km} = 1000 \text{ m} \triangleq 1000 \times 39 \cdot 37 \text{ in} = 39\ 370 \text{ in}$$

and

$$\tfrac{5}{8} \text{ mile} = \tfrac{5}{8} \times 1760 \text{ yd} = \tfrac{5}{8} \times 63\ 360 \text{ in} = 39\ 600 \text{ in}.$$

These are clearly very near to one another in length. The error is 230 in $\triangleq$ 19 ft.

i.e. **1 kilometre is approximately 19 ft short of $\frac{5}{8}$ mile.**

Although Ex. 1 above is interesting, it is important that the reader should practise the use of metric measurements *in their own right*, not trying to convert them always to the Imperial system. Such ideas as using centimetres and millimetres on a ruler, taking long strides of about 1 metre, and practising the use of km/h (kilometres per hour) instead of miles per hour on a car speedometer, all help. It is worth bearing in mind that 30 mile/h $\triangleq$ 50 km/h, for, from above,

$$50 \text{ km/h} \triangleq \tfrac{5}{8} \times 50 \text{ mile/h} = 31\tfrac{1}{4} \text{ mile/h.}$$

NOTES (1) The standard abbreviations for time will be h for hour and s for second.

(2) The solidus (/) is used instead of *per*, and leads to speeds being given in the form km/h, meaning 'kilometres per hour' (or kilometres÷hours), and m/s, meaning 'metres per second'; the word 'per' is retained when the words are written in full.

For those carrying out mathematical and scientific studies in greater depth, m s⁻¹ will be used instead of m/s; it means the same, for, from above and from page 46,

$$m/s = \frac{m}{s} = m\ s^{-1}.$$

Ex. 2. Express the speed of 90 km/h (90 km h⁻¹) in metres per second.

We have $\dfrac{90\ km}{1\ h} = \dfrac{90 \times 1000\ m}{3600\ s} = \mathbf{25\ m/s\ (m\ s^{-1})}.$

EXERCISE I

1. Express the following lengths in metres, putting down a zero in front of the decimal point where necessary:

(a) 2 km	(b) 3·75 km	(c) 41½ km	(d) ⅝ km
(e) 7 cm	(f) 38 cm	(g) 5·76 cm	(h) 650 cm
(i) 6 mm	(j) 82 mm	(k) 7 cm 6 mm	(l) 457 mm.

2. Using the table on p. 46 above, deduce the following approximations:

Imperial unit	Metric length
1 inch	2·540 cm
1 yard	0·9144 m
1 mile	1·609 km

3. In Britain there are sundry speed limits, including those of 30, 40, 50 and 70 mile/h. Taking 1 km ⌒ ⅝ mile, express these speeds (a) in km/h, (b) m/s.

4. A manufacturer who has been selling 60 yd bales of material at £72 a bale changes over to selling 50 m bales. How much should he charge, at the same rate?

5. Jones has a rectangular garden 45 ft wide and 120 ft long. The (45 ft) frontage on the road is bounded by a wall. If he needs new side and back fences, how many metres run of fence is needed?

6. A train of length 225 m is travelling at 90 km/h. How long will the train take to pass a signal?

4. Area

The fundamental unit of area is the *square metre*, which is the area of a square of side 1 metre. Its abbreviation is m².

Examples. $1 \, m^2 = 10^2 \, cm \times 10^2 \, cm = 10^4 \, cm^2$
$= 10 \, 000 \, cm^2.$

$1 \, km^2 = 10^3 \, m \times 10^3 \, m = 10^6 \, m^2$
$= 1 \, 000 \, 000 \, m^2.$

Consider now $\frac{1}{100} \, m = 1 \, cm$
i.e. $0 \cdot 01 \, m = 1 \, cm$

then $0 \cdot 01 \times 0 \cdot 01 \, m^2 = 1 \, cm^2$
i.e. $0 \cdot 0001 \, m^2 = 1 \, cm^2$

and so we can build up the table

$$100 \, m^2 = 1 \, Dm^2, \text{ also called } 1 \, are$$
$$10 \, 000 \, m^2 = 1 \, hm^2, \text{ also called } 1 \, hectare$$
$$1 \, 000 \, 000 \, m^2 = 1 \, km^2$$
$$0 \cdot 01 \, m^2 = 1 \, dm^2$$
$$0 \cdot 0001 \, m^2 = 1 \, cm^2$$
$$0 \cdot 000 \, 001 \, m^2 = 1 \, mm^2$$

Not all of these results are of much practical value.

In accordance with SI Units, area will be measured in the *square centimetre* (cm²), *square metre* (m²), *square kilometre* (km²); also, because of its convenient size for everyday life, the *hectare* (pronounced HEKTAIR) $\simeq 2\frac{1}{2}$ acre is likely to supersede the acre. The hectare (ha) is shown above and is 10 000 m²; it is 100 ares, where the are is 1 Dm².

SI unit	Area in square metres	Imperial equivalent
1 sq. centimetre	$0 \cdot 0001 \, m^2$	$0 \cdot 155 \, in^2$
1 sq. metre	$1 \, m^2$	$1 \cdot 196 \, yd^2$
1 hectare	$10 \, 000 \, m^2$	$2 \cdot 47$ acre
1 sq. kilometre	$1 \, 000 \, 000 \, m^2$	$0 \cdot 386 \, mile^2$

Ex. 3. How many tiles 6 cm × 9 cm are needed to cover a wall of length 3·6 metres to a height of 1·8 metres, if the tiles are laid longways?

In this kind of question there is always the problem that the tiles may not fit exactly. At this early stage, however, they have been 'adjusted' to fit!

$$\frac{\text{Length of wall}}{\text{Length of tile}} = \frac{3 \cdot 6 \, m}{9 \, cm} = \frac{360 \, cm}{9 \, cm} = \frac{360}{9} = 40$$

$$\frac{\text{Height of wall}}{\text{Height of tile}} = \frac{1 \cdot 8 \text{ m}}{6 \text{ cm}} = \frac{180}{6} = 30$$

∴ Number of tiles needed $40 \times 30 = \mathbf{1200}$
(i.e. 30 rows of 40 in each row).

Ex. 4. Taking one metre as 39·6 inches approximately, show that 1 hectare is very nearly $2\frac{1}{2}$ acres.

We have 1 hectare $= 10\,000 \text{ m}^2 = 10\,000 \times (39 \cdot 6)^2 \text{ in}^2$
  1 acre $\quad = 4840 \text{ yd}^2 = 4840 \times (36)^2 \text{ in}^2$

$$\therefore \frac{1 \text{ hectare}}{1 \text{ acre}} = \frac{10\,000}{4840} \times \left(\frac{39 \cdot 6}{36}\right)^2$$

$$\frac{10\,000}{4840} \times \frac{396}{360} \times \frac{396}{360} = 2\tfrac{1}{2}$$

∴ **1 hectare is approximately $2\frac{1}{2}$ acres.**

This is a bit naughty as 1 m $\frown$ 39·37 in, but 39·6 in cancels out very conveniently! Using the more accurate figure would have led to heavy arithmetic more suited to the use of logarithms (see later). The difference is not great, *c.f.* p. 48 which shows a more accurate result as 2·47 acres.

EXERCISE 2

1. Express the following areas in square metres, putting down a zero in front of the decimal point where necessary:

(a) 3 km²   (b) 1·45 km²   (c) 0·307 km²   (d) $\frac{3}{25}$ km²
(e) 5 dm²   (f) 512 cm²   (g) 3·62 cm²   (h) 7500 cm²
(i) 8 mm²   (j) 47 Dm²   (k) 824·6 cm².

2. Using the table on p. 48, deduce the following approximations:

Imperial unit	Metric equivalent
1 in²	6·45 cm²
1 ft²	929 cm²
1 yd²	0·836 m²
1 mile²	2·59 km²

3. A rectangular field is 80 metres long and 64 metres wide. What is its area in hectares?

4. Express one acre in square metres. (Take 1 m = 39·37 in.)

5. A fitted carept is bought in a roll, of width 75 cm. What is the length of carpet needed to cover the floor of a rectangular room 7 metres long and 4·5 metres wide? [*Hint.* Lay the carpet longways and, in this case, an exact number of widths is obtained.]

5. Volume and Capacity

The basic unit of volume is the *cubic metre*, which is the volume of a cube of side 1 metre. Its abbreviation is m³.

Example. 1 m³ = 10^2 cm × 10^2 cm × 10^2 cm
          = 10^6 cm³ = 1 000 000 cm³.

In SI units, the only quantities we are likely to use widely are the *cubic centimetre* (cm³), *litre* (l) = 1000 cm³ and *cubic metre* (m³).

It is interesting to note that

$$10^3 \text{ (i.e. 1000) cm}^3 = 1 \text{ l; } 10^3 \text{ (i.e. 1000) l} = 1 \text{ m}^3.$$

Hence the litre is the same as the cubic decimetre, although one will not often give it the latter name. For everyday life, the litre is the most useful of the three standard units listed above. In laboratory work, and sometimes in the kitchen, the cubic centimetre comes into its own. For the measurement of large quantities of materials, such as sand, the cubic metre is adopted.

SI unit	Volume in m³	Imperial equivalent
1 cubic centimetre	10^{-6} m³	0·061 in³
1 litre	10^{-3} m³	1·76 pint
1 cubic metre	1 m³	220 gallon

The cubic millimetre is so small that we should require a thousand million to make one cubic metre, for 1 m = 10^3 mm = 1000 mm,

∴ 1 m³ = 10^9 mm³ = 1 000 000 000 mm³.

It is unlikely that the cubic millimetre will have practical application in everyday life.

6. Mass. (In the past we referred to this, inaccurately, as WEIGHT.)

The *mass* of a body is the quantity of matter in it, dependent on the aggregate of its molecules, whereas the *weight* is the gravitational *force* exerted on the body by the Earth. It is well known that Sir

Isaac Newton (1642–1727) was interested in an apple which fell from a tree. Among the brilliant deductions he made from his subsequent investigations was a law which can lead to the equation

$$\text{Weight} = \text{Mass} \times \text{gravitational acceleration.}$$

The further a body is from the surface of the Earth, the smaller is the gravitational acceleration and, as the *mass* is constant, the above equation shows that the *weight* is smaller. The author's friend, Cyril Bolton, enlarged on this by suggesting that a housewife of the future, buying one kilogramme of potatoes to make chips, would get an awful lot of the vegetables to make up this *weight* as measured by a spring balance, if she were at her shop in a satellite orbiting the Earth at the time of purchase! The potatoes would be virtually weightless (although just as capable of satisfying one's inner needs) unless a spin were given to the satellite to create an artificial gravitational field.

It is best for children at school to move gradually towards the adoption of the word *mass*, but for adults not directly concerned with scientific or engineering work the distinction is not serious. Although it may well be frowned upon by those weaned on 'weight', we shall hereinafter adopt the word 'mass'.

The unit of mass (given in some detail among the TABLES, pp. xiii–xvi) is the *gramme* (g). It is possible to construct a table which corresponds in layout with those for length, area or volume:

$$
\begin{aligned}
1000\,\text{g} &= 1\ \textit{kilo}\text{gramme (kg)} \\
100\,\text{g} &= 1\ \textit{hecto}\text{gramme (hg)} \\
10\,\text{g} &= 1\ \textit{deka}\text{gramme (Dg)} \\
0{\cdot}1\,\text{g} &= 1\ \textit{deci}\text{gramme (dg)} \\
0{\cdot}01\,\text{g} &= 1\ \textit{centi}\text{gramme (cg)} \\
0{\cdot}001\,\text{g} &= 1\ \textit{milli}\text{gramme (mg).}
\end{aligned}
$$

In accordance with SI units, the only important ones are the *milli-gramme* (mg), *gramme* (g) and *kilogramme* (kg). We shall also use the *tonne* (metric ton) = 1000 kg.

The gramme is small, there being approximately 28·3 g = 1 oz. The milligramme is thus *very* small and is used in accurate laboratory measurements. The kilogramme will be widely used in purchasing goods by mass, e.g. 3·5 kg, for 1 kg ⌢ 2·206 lb. For weighing out small amounts in the kitchen, it is clear that the gramme will also have its uses. The tonne will be used for heavy commodities, e.g. coal, locomotives, ships' cargoes, etc.

$$1 \text{ tonne} = 1000 \text{ kg} \mathrel{⌢} 2205 \text{ lb.}$$

This is only 35 lb short of the former English ton (2240 lb).

SI unit	Mass in grammes	Imperial equivalent
1 milligramme	0·001 g	0·000 035 oz
1 gramme	1 g	0·0353 oz
1 kilogramme	1000 g	2·205 lb

There is a very useful relationship between volume and mass in the metric system. At 4°C (more accurately, the temperature is 3·98°C), one cubic centimetre of water has a mass of one gramme. Now $1 \text{ cm}^3 = \frac{1}{1000}$ litre (hence the cubic centimetre is also called the *millilitre*, when measuring liquids), and therefore

$$1 \text{ cm}^3 \text{ of water has a mass of } 1 \text{ g}$$
$$\therefore \quad 1000 \text{ cm}^3 \text{ ,, \quad ,, \quad ,, ,, ,, ,, } 1000 \text{ g}$$
$$\therefore \quad 1 \text{ litre \quad ,, \quad ,, \quad ,, ,, ,, ,, } 1 \text{ kg}$$

and 1 m^3 of water = 1000 litres of water has a mass of 1 tonne (= 1000 kg). In the above, we have ignored variations for modest changes of temperature.

Finally, there is a practical point. On the Continent, the housewife may buy her vegetables by the 'kilogramme', which is rather a long word. The official abbreviation 'kg' does not lend itself to speech and so she has adopted 'kilo', e.g. '2 kilos of potatoes'. It is not wise to *write* this abbreviation, however, as it merely means '1000'.

Ex. 5. Calculate the volume of metal in a rectangular sheet of length 3·15 m, breadth 24 cm and thickness 0·45 mm.

We convert all the lengths of the same units, say, centimetres.

$$\therefore \text{ Required volume} = 315 \times 24 \times 0·045 \text{ cm}^3$$
$$= \mathbf{340·2 \text{ cm}^3}.$$

Ex. 6. Find the mass, in grammes, of a pint of water, given that 1 litre $\triangleq$ 1·76 pint.

$$\text{Mass of one litre of water} = 1 \text{ kg} = 1000 \text{ g}$$

$$\therefore \quad \text{Mass of one pint} \triangleq \frac{1000}{1·76} \text{ g} \triangleq \mathbf{568 \text{ g}}.$$

EXERCISE 3

1. Write down the volume of a rectangular solid:
   (a) 6 m by 40 cm by 2 mm.
   (b) 45·5 m by 3.5 m by 1·6 mm.

2. What is the capacity, in litres, of a rectangular tank 1·2 m long, 70 cm wide and 45 cm high?

3. Given that 1 kg $\backsim$ 2·205 lb, find 1 lb in terms of a kilogramme.

4. Given that 1 gallon $\backsim$ 4·543 litres, show that 1 gallon of water has a mass of approximately 10 lb. [*Hint.* Use the result of No. 3 above.]

5. Wine is sold by the litre. How much is this in pints? (Take 1 gallon $\backsim$ 4·543 litres and give the answer correct to 3 significant figures.)

6. What is the volume of a brick 9 in by 4½ in by 3 in in cubic centimetres (correct to 3 significant figures)?

7. A reservoir holds 3 000 000 tonnes of water. How many cubic metres of water are there?

8. If a brick has a mass of 3·5 kg, how many bricks can a lorry carry if its payload is 4·75 tonnes? (Give the answer correct to the nearest 10 bricks.)

9. Assuming that 1 in $\backsim$ 2·54 cm, prove results (1) (2) (3) in the table below.

Imperial unit	Metric equivalent
1 in³	16·39 cm³
1 ft³	0·0283 m³
1 yd³	0·765 m³
1 gallon	4·543 litre

Assuming the result (4) is true above, deduce that 1 cu m $\backsim$ 220 gallon as stated earlier.

10. The tyres on a motor car are to be kept at a pressure of 25 lb/in². Express this in kilogrammes per square centimetre (kg/cm²).

11. A rolling mill purchases 240 tonnes of a metal which has a mass of 7·2 tonnes per cubic metre. The metal is rolled out into a sheet of breadth 2 m and thickness 2·5 mm. What length of metal is produced?

CHAPTER SEVEN

MONEY

1. The British Coinage

The silver penny (Latin *denarius*) was the d of our £ s d system, which became obsolete on 15th February, 1971. It came into general use in Anglo-Saxon times being used by various potentates, e.g. *c* 765 Heabert[6], King of Kent; *c.* 765 Jaenberht, Archbishop of Canterbury; *c.* 757 Offa[7], King of Mercia (famed for his dyke, still largely extant and separating Wales from England); *c.* 760 Beonna, King of East Anglia. The silver penny remained the principal coin for a long time, until, in fact, gold money was also introduced in 1344, *temp.* Edward III. The first important gold coin was, strangely enough, a florin (equivalent to 6s), but quite different from the florins we know (now replaced by 10p pieces) and which date from Victoria's reign. A florin of 1344 is currently valued, in reasonable condition, at about £8000, so it will be appreciated that few specimens have survived.

The coins from Edward III's reign onwards became much more varied, e.g. half-florin (leopard) and quarter-florin (helm). Then, in 1344–6, the noble (6s 8d) appeared, together with sub-divisions of it. Again, in 1344, we come across the half-penny and farthing (both in silver) and, in 1351, the groat (4d) and half-groat. In the time of Edward IV, there appeared, in 1464, the ryal (rose-noble, a gold coin of 10s), and its sub-divisions of a half and a quarter; contemporarily with this, it appears that the angel (6s 8d) superseded the noble. With the ascent of Henry VII, in 1485, we find the gold sovereign of 20s and the testoon (1s). Under Henry VIII's despotic rule some odd things happened to the coinage. This was partly because, in 1526, some coins were manufactured in 22 carat gold and later, in 1543, there was a substantial debasement of money. The points of interest are in the table opposite.

By 1544 we come across the half-sovereign (10s), the short lived George-noble and its half having been discontinued. What a muddle! Throughout all this, the groat, half-groat, penny, halfpenny and farthing (all in silver) continued to prosper; the testoon was also still minted, but by 1549, *temp.* Edward VI, it was replaced by the shilling. Under Mary I's reform, 1553, a 'fine' sovereign (of 30s) and a ryal

[6] It appears that only *one* of Heabert's silver pennies has survived.

[7] Two unique *gold* coins of Offa exist; we then meet few other gold coins for a long time.

1509–26	Sovereign	20s
	Angel	6s 8d
	Half-Angel	3s 4d
1526–44	Sovereign	22s 6d
	Angel	7s 6d
	Half-Angel	3s 9d
and then	George-Noble	6s 8d
	Half-George-Noble	3s 4d
	Crown of the Rose	4s 6d
	Crown of the Double Rose	5s 0d
and we meet	Half-Crown[8]	2s 6d

(of 15s), together with an angel (now 10s) and half-angel, were the gold coins. These continued in Elizabeth I's day; at the same time, one finds the silver three-pence, three half-pence and rather rare three farthings; thus, in *silver*, there were 5s, 2s 6d, 1s, 6d, 4d, 3d, 2d, 1½d, 1d, ¾d, ½d!

With the ascent of James I (James VI of Scotland) in 1603, the sovereign reverted to 20s, but the second coinage of this monarch included some little-known items, such as the rose-ryal (30s), spur-ryal (15s), angel (10s), half-angel (5s 6d), unite (20s), double crown, Britain crown, thistle-crown (4s), and even a laurel (20s) and its half and quarter. Chaos continued in Charles I's time, all sorts of coins being minted in different isolated towns during the Civil War. During the Commonwealth, Cromwell used a gold unite, double crown, crown, half-crown, shilling, sixpence, half-groat, penny and half-penny (silver) and even in 1656, a 50s piece.

With the Restoration in 1660 we come across the guinea, introduced as follows: 1663 guinea; 1664 two guinea piece; 1668 five guinea piece; 1669 half-guinea.

Base metal coins are to be found dating from 1672, in the form of a copper half-penny and farthing, and in 1684–5, a tin farthing. The coinage now steadied for more than a century and, ignoring local penny and half-penny tokens, stood as:

Gold	Five guineas, two guineas, one guinea, half guinea (and, more rarely, a quarter-guinea)
Silver	Crown, Half-crown, shilling, sixpence, four-pence, threepence, twopence, penny
Base metal	Halfpenny, farthing

[8] Presumably Half Crown of the Double Rose, to give it its full title.

In 1797, Matthew Boulton (1728–1809), engineer, manufactured an enormous 'cartwheel' copper 2d and a correspondingly massive 1d, but there was such an outcry from the public about the mass of these coins that they were quickly withdrawn.

The last guineas were minted in 1813, *temp.* George III. In 1816, a new gold coinage, five pound, two pounds and one pound (sovereign) superseded the guinea (most carefully preserved on paper by ladies' dress shops and other establishments, as a means of presenting apparently attractive prices to customers, until 1970!).

At last, in 1825, during George IV's reign we find a moderate-sized copper penny.

The silver 4d, 3d, 2d, 1d were still issued as Maundy money, even until the 1970's, but a silver groat (4d) and threepence of different design were used in circulation, the former until 1855 (ignoring those for colonial use) and the latter until 1944, George VI's reign.

The government, in 1849, decided that the time had come for decimalisation of the currency and introduced the florin (2s); inscribed 'One florin—One tenth of a pound'. The half-crown was discontinued in 1850, but the public would have none of this, and by 1874 the half-crown was again minted. It would appear that the government was equally reluctant to abandon its florin and so the two coins were minted side by side until 1967. We had now arrived at the well-known coinage.

Gold	£5, £2 (only issued at rare intervals) £1 (last minted for *circulation* in Britain in 1918, although for commemorative purposes and for collectors they have been minted as late as Elizabeth II's reign).
Silver	5s (discontinued in circulation 1902; minted for commemorative purposes from time to time until 1965) 2s 6d; 2s; 1s; 6d (92·5% silver to 1920; 50% silver 1920–46; cupro-nickel since then; 2s 6d withdrawn 1969–70).
Silver or Brass	3d ((*a*) material used as in 2s 6d, withdrawn 1944; (*b*) in brass 1937–67).
Bronze	1d, ½d, ¼d (withdrawn as follows: ¼d, 1955; ½d, 1969; 1d, 15th February 1971).

The 2s (florin) and 1s are replaced by the new 10 penny piece (10p) and new 5 penny piece (5p) respectively, the new coins being of identical size to their predecessors. The sixpence, because of public demand, has been temporarily maintained in the form of 2½p.

Thus we arrive at our new coinage:

£1 (as before); 50p (=10s); 10p (=2s); (5p (=1s); 2½p (=6d); 2p (=4·8d); 1p (=2·4d); ½p (=1·2d).

The 50p, 10p, 5p, 2½p are of cupro-nickel and the smaller coins are of bronze.

The purpose of our decimal currency (100p = £1) is to avoid fractions (although the introduction of the ½p produces complications in this respect).

2. Use of the New Coinage

There are certain rules[9] which are recommended officially in order to save mistakes in using the new coinage. Although these rules are straightforward, there are a few pitfalls to be avoided. These are emphasised in the list below:

(1) The £ sign is used for pounds (sterling), as before: thus, £24.

(2) The sign for new pence is p and it is placed after the amount: e.g. 47p.

(3) The new halfpenny is to be shown as a vulgar fraction ½, as in 29½p.

(4) The decimal point is used to separate pounds from new pence. As there are 100p in £1, two figures must be given after the decimal point (unless the sum of money concerned is a whole number of pounds), e.g.

£3·45 means £3 and 45p
£8·07 means £8 and 7p.

If the zero were omitted in the £8·07 above, it would be possible to read the amount incorrectly as £8·7 = £8 and 70p.

(5) In writing the new coinage, *either* the pound sign (£) *or* the pence sign (p) is used, but *both signs should not be used together*, e.g.

(a) £28·46½, (b) £4·509·08, (c) 53p, (d) 1½p.

Note that it would be quite correct to write (c) above as £0·53 and (d) as £0·01½.

There is no reason why, in calculations, the example (a) should not be written as 2846½p if it is helpful, but the answer should certainly be given as £28·46½. It is then easier to visualise the sum involved.

[9] Adapted from *Decimal Currency*, issued by HMSO.

(6) Whenever a sum of money is not exact, it is adjusted to the nearest $\frac{1}{2}$p. This rule is tiresome. Consider the cost of 300 gm cheese at $47\frac{1}{2}$p per kg. We have the cost given by

$$\frac{300}{1000} \times 47\frac{1}{2}\text{p} = \frac{3}{10} \times \frac{95\text{p}}{2} = \frac{57\text{p}}{4} = 14\frac{1}{4}\text{p}$$

(which could be 14p or $14\frac{1}{2}$p to the nearest $\frac{1}{2}$p).

Thus, for several items on a bill we might get some awkward calculations. The reason for this lies in the fact that 240 has many more factors than 100.

Our old system was £1 = 240d, where the factors of 240 are 1, 2, 3, 4, 5, 6, 8, 10, 12, 15, 16, 20, 24, 30, 40, 48, 60, 80, 120, 240.

Altogether there are 20 different factors.

Now consider our new system, £1 = 100p, in which the factors are:

$$1, 2, 4, 5, 10, 20, 25, 50, 100.$$

There are now only 9 factors.

The idea was that, in using decimals instead of fractions, full compensation would be made for the loss of factors, and hence of aliquot parts (i.e. parts contained a whole number of times in a whole thing). It would have worked well had we not been compelled to use the $\frac{1}{2}$p, because we retained so large a unit as £1 instead of a smaller 10s unit, which we could have designated a dollar ($1), where 1p would have been of size akin to the former 1d.

(7) The coins in use are $\frac{1}{2}$p, 1p, 2p (bronze); $2\frac{1}{2}$p, 5p, 10p, 50p (cupro-nickel). The pound note is continued. It is curious that the $2\frac{1}{2}$p unit is still to be found in circulation, for it only exists in the form of the old 'sixpence'.

(8) The guinea is finally abolished (about 160 years after the last such coin was manufactured). Thus a price of 63 gns will have to be written as £66·15. This makes the true cost much clearer to a buyer. The example also nicely illustrates the purpose of rule (7) above during the transitional period.

$$\begin{aligned}
63 \text{ gns} &= \text{£63 plus 63s, i.e. £63 plus 63 'fives'}\\
\text{Now 63 fives} &= 63 \times 5\text{p} &= 315\text{p} = \text{£3·15}\\
\therefore \quad 63 \text{ gns} &= \text{£}(63 + 3·15) = \text{£66·15}.
\end{aligned}$$

(9) The procedure to be adopted in writing cheques is illustrated by the four examples below, which cover the possible cases. On a cheque, a hyphen is written instead of a decimal point. This is merely for the sake of clarity. After all, an indistinct decimal point could easily be missed!

Note that the £ sign must appear in the Figures section and that at least *one* number (zero if necessary) must appear between the £ sign and the hyphen. In printing we will, as already explained, find (say) £37·00, £5·08, £20·70 or £0·46 to correspond with the above four amounts. It may be desirable to alter typewriters to carry out the

Hand-written	Figures
Thirty Seven Pounds	£37-00
Five Pounds 08	£5-08
Twenty Pounds 70	£20-70
Forty Six pence	£0-46

same placing of the decimal point, but, until this is done, we must accept £37.00, £5.08, £20.70 or £0.46 as correct, *but only when type-written.*

Ex. 1. Mrs. Green purchases items costing 47p, 9½p, 62½p and £3. What is her total bill?

The layout is as for decimals (Ch. 5, Section 2), in that we use the decimal point as our guide in putting down numbers to be added or subtracted in columns. There is a slight modification to take into account the ½p.

£		
	47	←*Either* £ sign *or* p can be used by itself in the correct place
	09½	←Note that the zero is essential
	62½	
3	00	←We also put the zeros in for safety.
4	**19**	**The total cost is £4·19.**

The problem could alternatively have been written as

$$£$$
$$0·47$$
$$0·09½$$
$$0·62½$$
$$3·00$$
$$\overline{£4·19}$$

The advantage of the first layout is that the difficulty some people have in putting down decimal points neatly in a vertical line, when shopping in a hurry, is avoided!

Ex. 2. Smith buys 200 g of ham at £2·16 a kg and 400 g of cheese at £1·74 a kg. How much change will he have from a £5 note?

Ham $\dfrac{200}{1000} \times 216\text{p} = \dfrac{216}{5}\text{p} = 43\cdot2\text{p} \backsimeq 43\text{p}$

Cheese $\dfrac{400}{1000} \times 174\text{p} = \dfrac{348}{5}\text{p} = 69\cdot6\text{p} \backsimeq 69\tfrac{1}{2}\text{p}$

Total cost £1·12½

Change received (£5—£1·12½) = **£3·87½**

Add together the following:

1. 20p, 37½p, 4½p

2. 61½p, £4·05, 9½p

3. £169·08, £84, £0·94

4. 65½p, 82½p, 47p.

Subtract the following:

5.   £
  63·04
  47·97

6.   £
  16·84
  15·86½

7.   p
  47
  38½

8. Subtract the sum of 74½p and 37p from £5.

9. Find the total cost of 1¼ kg potatoes at 15½p a kg and 750 g pears at 68p a kg.

10. Find the cost of 3½ kg tomatoes at 72p a kg, 5 kg potatoes at 16½p a kg and 1½ kg carrots at 19½p a kg. How much change will there be from a £5 note? (Answer to the nearest 1p.)

3. Multiplication and Division of Money

The processes are simple and very largely follow the lines shown in Chapter 5. There is no doubt whatsoever that multiplication and division of decimal currency is easier than with the former £ s d.

Ex. 3. Find the cost of 47 articles at 73½p each.

p
73½
47 ×
————
514½
294
————
3454½p

The reader will observe that, inevitably, the ½p has to be used with caution: $40 \times \tfrac{1}{2} = 20$, giving 2 to be carried in the 'tens' column.

Total cost £34·54½.

Alternatively, we could have laid this out rather more clearly, as follows,

£	
	73½
	47 ×
5	14½
29	40
34	54½

Ex. 4. Find the number of articles which can be bought for £250, if each costs 73½p. How much change will there be?

$$\frac{£250}{73\frac{1}{2}\text{p}} = \frac{25\ 000\text{p}}{73\frac{1}{2}\text{p}} = \frac{25\ 000}{73\frac{1}{2}} = \frac{50\ 000}{147},$$

on converting the top line to pence, and then removing the p sign top and bottom, and finally doubling top and bottom to get rid of the ½.

$$
\begin{array}{r}
340 \\
147\overline{)50\ 000} \\
44\ 1 \\
\hline
5\ 90 \\
5\ 88 \\
\hline
20 \quad \text{Remainder}
\end{array}
$$

Thus 340 articles can be bought.

The *remainder*, 20, poses a problem. We must remember that we *doubled* the top and bottom lines of the fraction, so we must now *halve* the remainder to get the change, in pence. It is therefore **10p.**

EXERCISE 2

Multiply (Nos. 1–4):

1. £3·40 by 17   **2.** £0·54 by 28   **3.** 37p by 43   **4.** 65½p by 129.

Divide, giving the answer correct to the nearest 1p where necessary (Nos. 5–10):

5. £572 by 11   **6.** £3007 by 31   **7.** £84·50 by 133

8. £64·28 by 190   **9.** £2447 by 392   **10.** £251 by 493.

11. Find the cost of 73 tonnes of coal at £23·45 a tonne.

12. How much will 509 articles be, at 47p each?

13. If stacking chairs cost £27·60 a dozen, how much will a school have to pay for 614 chairs?

14. Pencils are priced at £7·56 a gross. What will be the cost for 1000 pencils?

15. An ironmonger wishes to purchase garden shears at £4·55 each. How many will he get for £70 and how much change will there be?

16. Find, in lowest terms, what fraction 67½p is of £7·65.

17. If electric torches cost 31½p each, how many can be bought for £17 and how much change will be left?

18. Jones dies and leaves £6874. He directs that the money shall be divided among his wife and three children, so that his wife receives four times as much as each child. The children are to get equal shares. How much is left to the wife?

19. Tompkins took a party of 29 children to the theatre. The seats cost 42½p each. How much change did he have from £15?

20. What is the cost, in £, of motoring 17 952 km at 5½p a kilometre?

4. Delay in the Completion of Conversion from the Imperial System to Metric Weights and Measures

The delay in conversion to metric measurements fully justifies the inclusion of some of the simpler tables of Imperial weights and measures, probably for the next few years, so that readers who have not learned them at school will have them to hand.

LENGTH

12 inches (in) = 1 foot (ft)
3 feet (ft) = 1 yard (yd)
22 yards (yd) = 1 chain (ch)
10 chains (ch) = 1 furlong (fur)
8 furlongs (fur) = 1 mile (mi)

Hence, 220 yd = 1 fur
5280 ft = 1760 yd = 1 mi
At sea, the following are used
6 ft = 1 fathom
200 yd = 1 cable

AREA

144 sq in = 1 sq ft
9 sq ft = 1 sq yd

4840 sq yd = 1 acre (ac)
10 sq ch = 1 acre (ac)
640 ac = 1 sq mile

VOLUME

1728 cu in = 1 cu ft

27 cu ft = 1 cu yd

MASS (AVOIRDUPOIS)

16 ounces (oz) = 1 pound (lb)
14 lb = 1 stone (st)
28 lb = 1 quarter (qr)

4 qr = 1 hundredweight (cwt)
20 cwt = 1 ton (t)
2240 lb = 1 t

CAPACITY*

4 gills = 1 pint (pt)   2 gall = 1 peck (pk)
2 pt = 1 quart (qt)   4 pk = 1 bushel (bush)
4 qt = 1 gallon (gall)   8 bush = 1 quarter (qr)

*Note: the table is easy to remember: 4, 2, 4, 2, 4, 8.
The measurement of time is *not* affected by metrication. It is based
largely on the sexagesimal scale (the scale of 60—used more than 2500
years ago by the Babylonians).

Ex. 5. Convert 2 mi 3 fur 118 yd to feet.

$$2 \times 5280 = 10560$$
$$3 \times 3 \times 220 = 1980$$
$$118 \times 3 = 354$$

Total distance   12894 ft

EXERCISE 3

1. Convert 249 831 inches to miles and yards, correct to the nearest yard.
2. Express 7 ton 6 cwt 76 lb as (a) pounds, (b) kilogrammes, to the
   nearest kilogramme.
3. Find in tons the mass of 1 000 000 oz, correct to 2 decimal places.
4. Express 363 000 sq yd in (a) acres, (b) hectares, correct to 2 decimal
   places.
5. Wine is sold in 70 cl, 75 cl and 1 l bottles. Express these volumes as
   decimals of one pint. (*Note* cl = centilitre; l = litre).

†5. Practice

Practice has had many applications in arithmetic. It consists of
expressing a quantity as a set of aliquot parts. An aliquot part can
be considered as an exact divisor of a number.

When our units of length, area, volume and mass are fully metricated, as well as our money, it is unlikely that there will be much
purpose in using the method of practice.† *This section is therefore
included for use during the transitional period.*

Consider, for example, 5 ton 6 cwt 1 qr. We can put it in aliquot
parts thus:

5 ton = 5 × 1 ton   Each line is an exact divisor of the line
5 cwt = ¼ × 1 ton   (except for lines 1 and 2 which are easily
1 cwt = ⅕ × 5 cwt   understood).
1 qr = ¼ × 1 cwt

Ex. 6. Find the cost, to the nearest 1p, of 5 ton 6 cwt 1 qr of coal at £18·65 a ton.

		£	
	Cost of 1 ton	18	65
∴	Cost of 5 ton	93	25
	,, ,, 5 cwt = $\frac{1}{4}$ of 1 ton	4	66 25
	,, ,, 1 cwt = $\frac{1}{5}$ of 5 cwt		93 25
	,, ,, 1 qr = $\frac{1}{4}$ of 1 cwt		23 31
	Total cost **£99·08.**	£99	07 81

EXERCISE 4

1. Find the cost of 27 jars of marmalade at 29$\frac{1}{2}$p a jar.

2. Find the cost of 8$\frac{1}{4}$ lb of bacon at 94$\frac{1}{2}$p a lb.

3. What is the cost of 7 yd 2 ft 4 in of material at £1·38 a yard?

4. How much would 274 yd of cable cost at 6$\frac{1}{2}$p a foot?

5. A farmer buys the crop growing on a field at £7·40 an acre. How much would it cost him to buy the crop on a field of area 17 acres 3 roods? (Remember that there are 4 roods to the acre.)

6. Find the cost of 8 ton 7 cwt 3 qr coal at £19·82 a ton. (To the nearest 1p.)

7. An insurance agent is paid 17$\frac{1}{2}$p in every £ which he gets in orders for the manufacture of cosmetics. How much will he be paid for orders of £287?

6. Unitary Method

This method, which is often applied to money problems, is helpful to those who initially have difficulties with the ideas of ratios, given in Chapter 8, q.v. As the name indicates, we reduce the cost to that of one article. An example will clarify the idea.

Ex. 7. If a club pays £165 for 132 theatre tickets, how much will it pay for 183 at the same price per ticket?

$$132 \text{ tickets cost } £165$$
$$\therefore \quad 1 \text{ ticket costs } £\tfrac{165}{132}$$
$$\text{and so} \quad 183 \text{ tickets cost } £\tfrac{165}{132} \times 183$$
$$= \textbf{£228·75.}$$

Note that: (1) we delay simplification until the end, (2) using ratios (Ch. 8), we get the same result, but laid out as £165 × $\frac{183}{132}$.

EXERCISE 5

1. If 4 plates cost 90p altogether, how much will 14 plates be?

2. Oranges are sold at 5 for 36p. How much, to the nearest 1p, will 8 oranges cost?

3. Find whether it is cheaper to buy 18 identical articles for 25p at one shop or 29 of the same articles for 38p at another shop.

CHAPTER EIGHT

RATIO, PROPORTION AND AVERAGES

1. Ratio

Suppose Mr. and Mrs. Barnett wish to buy a house and they see two which appeal to them. The first costs £14 400 and the second costs £16 800. We say that the *ratio* of the cost price of the first house to that of the second house is as £14 400 is to £16 800. This can be written:

$$\frac{\text{Cost price of first house}}{\text{Cost price of second house}} = \frac{14\ 400}{16\ 800} = \frac{6}{7}, \textit{ in lowest terms.}$$

Another way of writing this is 'the ratio of the prices is 6:7'. It is important that the numbers are put in the correct order. It will be seen that the £ signs on the right have been removed. This can always be done if the quantities, top and bottom, are the same. We have already used this idea in Chapter 4.

NOTE '6:7' is read '*as six is to seven*'.

Ex. 1. Osborn works 40 hours a week. Find the ratio of the time he works to the time when he is not working.

Number of hours in a week
$$\text{Time working} = 40\,\text{h}$$
$$\text{Number of hours in a week} = 7 \times 24 = 168\,\text{h}$$
$$\therefore \text{Time not working} = 168 - 40 = 128\,\text{h}$$
$$\therefore \text{Required ratio} = 40:128$$
$$= \textbf{5:16.}$$

Ex. 2. Apples have been increased in price from $61\frac{1}{2}$p to 69p per kilogramme. In what ratio has the price been increased?
The ratio in which price has been increased

$$= \frac{\text{New Price}}{\text{Old Price}}$$

$$= \frac{69\text{p}}{61\frac{1}{2}\text{p}} = \frac{46}{41}.$$

In a case of this kind where the ratio is of two fairly large numbers a better comparison is made by expressing the result in the form

x:1, i.e. making the new denominator 1, by dividing out by the original one.

∴ Ratio
New Price : Old Price ≏ 1·12 : 1.

$$41)46(1·12$$
$$\underline{41}$$
$$50$$
$$\underline{41}$$
$$90$$

Ex. 3. A photograph 12 cm by 9 cm is to be enlarged to fit a frame of longer side 16 cm. What length is the shorter side of the frame? What is the ratio of the new area of the picture to the original area, if the shape is unchanged?

$$\frac{\text{Shorter side of frame}}{\text{Longer side of frame}} = \frac{\text{Shorter side of photograph}}{\text{Longer side of photograph}}$$

$$\therefore \frac{\text{Shorter side}}{16} = \frac{9}{12}$$

i.e. Shorter side $= \frac{9}{12} \times 16 = $ **12 cm.**

$$\therefore \frac{\text{New area}}{\text{Old area}} = \frac{16 \times 12}{12 \times 9} = \frac{16}{9}$$

∴ **Ratio of the new area to the original area is 16 : 9.**

EXERCISE 1

1. Find the ratio of 165 m to 2·75 km.

2. What is the ratio of £1·19 to 73½p, in the form x:1?

3. It costs £11·25 a day to stay in a hotel in September, and £88·25 a week in August. By comparing the cost of a week's holiday in September with that of a week's holiday in August find the ratio of the costs, in lowest terms.

4. Baker earns £10·20 a day in a five-day week, and Hughes earns £68 a week. Find the ratio of their earnings. (Compare the amount earned by the men either in one day *or* in one week.)

5. During a strike, two-fifths of Smith's wages were lost. In what ratio was his income reduced?

6. During the week of a furniture sale, all furniture was reduced in the same ratio. If a piano originally costing £245 was offered at £182 what would be the sale price of a coffee table originally costing £17·50?

7. A man pays 42½p tax on every £1 he gets. What is the ratio of his net income (after paying tax) to the tax he pays (give the answer in the form n:1, correct to 3 sig. figs.)?

8. John has a bar of chocolate and he gives three-sevenths of it to his younger sister. What is the ratio of the amount he now has to the amount given away?

9. Express 66·5 litres as an exact decimal of 152 litres.

10. The ratio of a quantity A is to B as 4:5, and B is to C as 2:3. Find C if A is 32.

11. The profits of a firm totalled £7236 in 1970. If these profits are to be divided among three partners in the ratio 4:3:2, find how much each receives.

12. A company manufactured cosmetics which it marketed at twice the cost of production in 1970. During 1971 the cost increased in the ratio of 4:3 whilst the selling price was increased only in the ratio of 5:4. Calculate the ratio of the 1971 profit to the 1970 profit on each article.

2. Variation and Proportion

(a) **Direct Variation.** When two quantities are so related that their ratio remains constant, then either is said to *vary directly* as the other. If one quantity is y and the other is x, the ratio $\dfrac{y}{x} = k$ (a constant), i.e. $\mathbf{y = kx}$, so if $x = 1, y = k; x = 2, y = 2k$ and so on.

As an illustration, consider speed (in km/h). It is defined as the rate of change of distance, or

$$\text{Speed (km/h)} = \frac{\text{Distance (kilometres)}}{\text{Time (hours)}}.$$

If the speed is v km/h, distance s kilometres and time t hours, we have

$$v = \frac{s}{t}$$

$\therefore s = vt$ on multiplying both sides by t.

Suppose a car is travelling at 48 km/h

$$\text{then } s = 48t.$$

In 10 minutes the car travels $48 \times \frac{10}{60} = 8$ km
20 ,, ,, ,, ,, $48 \times \frac{20}{60} = 16$,,
30 ,, ,, ,, ,, $48 \times \frac{30}{60} = 24$,, etc.

If we draw a graph[10] with minutes plotted on the horizontal axis and kilometres on the vertical axis, we find the points obtained by marking off 10 minutes horizontally and 8 kilometres vertically (i.e.

[10] Graphs are explained in detail in Chapter 10.

the point A whose *coordinates* are (10, 8); 20 minutes horizontally and 16 kilometres vertically (i.e. the point B whose coordinates are (20, 16)); and so on. We get a series of points A, B, C . . . lying on a straight line passing through the origin O.

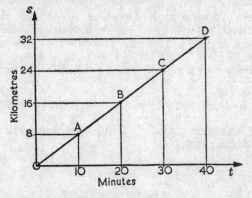

(b) **Inverse Variation.** Definition. The *reciprocal* of a number x is one (i.e. *unity*) divided by x, i.e. $\frac{1}{x}$.

When two quantities are so related that one of them varies directly as the *reciprocal* of the other, the numbers are said to *vary inversely*.

If one quantity is y and the other is x, then y varies as $\frac{1}{x}$, i.e. $y = k \times \frac{1}{x}$ (where k is a constant), or more briefly

$$y = \frac{k}{x}.$$

If $x = 1, y = k$; if $x = 2, y = \frac{1}{2}k$; if $x = 3, y = \frac{1}{3}k$ and so on. Again consider the equation

$$\text{Speed (km/h)} = \frac{\text{Distance (kilometres)}}{\text{Time (hours)}}.$$

Using the same notation as before

$$v = \frac{s}{t}.$$

Suppose a car has to travel a definite distance, say 240 kilometres, then

$$v = \frac{240}{t}.$$

If the time taken were 2 hr., $v = \frac{240}{2} = 120$ km/h

,, ,, ,, ,, ,, 3 ,, $v = \frac{240}{3} = 80$,,

,, ,, ,, ,, ,, 4 ,, $v = \frac{240}{4} = 60$,,

,, ,, ,, ,, ,, 5 ,, $v = \frac{240}{5} = 48$,, etc.

If we were to draw a graph of speed against time, i.e. time (say in hours) horizontally and speed (in km/h) vertically we would no longer have a straight line, as we see by plotting the points (2, 120), (3, 80), (4, 60) etc.

(The curve we get is actually called a rectangular hyperbola.) (*See below.*)

Had we, however, plotted a graph of $\frac{1}{v}$ against t we *would* have obtained a straight-line graph. We could have forecast this, because, taking our present example,

$$v = \frac{240}{t}$$

$$\therefore vt = 240$$

$$\text{i.e. } t = \frac{240}{v} = 240 \times \frac{1}{v}.$$

Putting $\frac{1}{v} = w$, then $t = 240w$, which is an equation of direct variation connecting t and w, and as we saw in section 2(*a*) above, this will yield a straight-line graph.

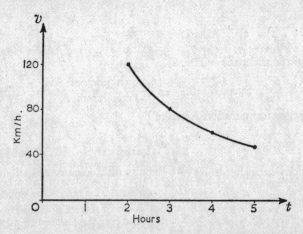

Ex. 4. Jones drives $115\frac{1}{2}$ km in 1 hr 52 min. How long would it take him to drive 132 km at the same speed?

(This is an example of direct variation: the greater the distance the greater the time.)

If the distance is increased in the ratio $132:115\frac{1}{2}$, then the time is increased in the same ratio; but the time was originally 112 min.

$$\therefore \text{New time} = 112 \times \frac{132}{115\frac{1}{2}} = \frac{112 \times 264}{231} \text{ min}$$

$$= 128 \text{ min} = \mathbf{2\ h\ 8\ min.}$$

Ex. 5. If 6 men take 20 days to paint a row of houses, how long would 8 men take?

(This is an example of inverse variation: the greater the number of men the shorter the time taken.)

If the number of men is *increased* in the ratio 8:6, the number of days is *decreased* in the ratio 6:8, i.e. it is multiplied by $\frac{6}{8}$.

$$\therefore \text{Number of days} = 20 \times \frac{6}{8}$$
$$= \mathbf{15.}$$

When the problem takes a more elaborate form, it is often useful to invent compound units. If 5 men are employed for 6 days they will have done 30 man-days of work.

Ex. 6. It costs £138 to hire 3 lorries for 5 days. How much could it cost to hire 7 lorries for 8 days?

Cost of hiring 3 lorries for 5 days is 15 lorry-days

,, ,, ,, 7 ,, ,, 8 ,, ,, 56 ,,

$\therefore$ Cost has increased in the ratio 56:15

$$\therefore \text{New cost} = £138 \times \frac{56}{15}$$
$$= £\frac{7728}{15}$$
$$= \mathbf{£515 \cdot 20.}$$

EXERCISE 2

1. The cost of a holiday is £56·70 a week. How much would it cost for 12 days?

2. If the average speed of a train is increased from 90 km/h to 100 km/h for a journey of 240 km, find how much time is saved?

3. Three partners would have to provide £3200 each to buy a business. What would be the cost to be borne by each, if there were 5 partners?

4. An expedition is equipped with provisions for 28 days. How long would the food last if each member had his ratio reduced by a quarter?

5. If 3 men are paid a total of £144·90 for 4 days' building construction, how much would it cost to employ 8 men for 7 days?

6. Williams & Co. have undertaken to build a road 2 miles long in 12 weeks. They employ 69 men. After 10 weeks 1½ miles have been constructed. How many more men must now be employed to complete the work on schedule?

3. Proportional Parts

If a quantity is divided into two or more parts so that there are a units in the first part, b in the second, c in the third, and so on, we say that the quantity is divided in the ratio $a:b:c$, etc.

For example, if £150 is divided into three parts in the ratio 2:3:5, then the first part contains 2 shares, the second part contains 3 shares and the third part contains 5 shares. Altogether there are $2+3+5$ (= 10) shares.

∴ The first part contains $\frac{2}{10}$ of the whole, i.e. $£150 \times \frac{2}{10} = £30$⎫
∴ The second part contains $\frac{3}{10}$ of the whole, i.e. $£150 \times \frac{3}{10} = £45$⎬
∴ The third part contains $\frac{5}{10}$ of the whole, i.e. $£150 \times \frac{5}{10} = £75$⎭

Ex. 7. A legacy is to be divided among May, John and Hilda in the ratio of 4:3:2. If the legacy is £720, find how much each receives.
Altogether there are $4+3+2$ shares, i.e. 9 shares.

∴ May receives $£\frac{4}{9} \times 720 =$ **£320**
   John receives $£\frac{3}{9} \times 720 =$ **£240**
   Hilda receives $£\frac{2}{9} \times 720 =$ **£160**

(Notice that $\frac{4}{9} + \frac{3}{9} + \frac{2}{9} = 1$, i.e. one whole legacy.)

EXERCISE 3

1. Divide £50 among 4 people in the ratio 6:5:3:1.

2. Three men and 2 women are employed. The earnings of a man as to that of a woman is 4:3. If the total amount earned is £450, find how much each receives. (There are 3 men each with 4 shares, and 2 women each with 3 shares.)

3. Three farmers, Jones, Wright and Siddons, share grazing land for their cattle. Jones and Wright each have 75 head of cattle and Siddons has 90. Jones removes his cattle after 2 months, but both Wright and Siddons graze theirs for 3 months. If the total rent is £129, calculate how much each should pay.

4. Find which town has the greater proportional rate of increase of population: Town A (1976)—217 800, (1977)—222 300; Town B (1976)—70 500, (1977)—71 900.

5. A tap can fill a bath with water in 5 minutes. The bath is emptied in 7 minutes when the plug is removed. How long would it take to fill the bath when the tap is on and the plug is removed?

6. What is the ratio of $4\frac{1}{4}$ to $2\frac{1}{4}$? Find the cost of $4\frac{1}{4}$ tonnes of coal if $2\frac{1}{4}$ tonnes cost £40·50.

4. Averages

When an experiment is repeated a number of times, it is of value to know the average result. For example, a part-time gardener may earn £30, £28, £37, £33, £36, £24, £29, £35 in 8 successive weeks. In order to assess his average income during that time it is only necessary to add these profits and to divide the result by the number of items added.

$$\therefore \text{ Average income} = \tfrac{1}{8}\{30+28+37+33+36+24+29+35\}$$
$$= \tfrac{1}{8}\times252 = \textbf{£31·50.}$$

That the average gives the *best* estimate of a number of events of this type is shown in books on statistics.

Sometimes an item may be repeated a number of times. The principle remains unchanged.

Ex. 8. 7 kg of tea at £2·47$\frac{1}{2}$ per kg are blended with 3 kg of tea at £2·70 per kg. At what price should the mixture be valued?

There are 10 kg altogether.

$$\therefore \text{ Average price} = \tfrac{1}{10}\{7\times247\tfrac{1}{2}p+3\times270p\}$$
$$= \tfrac{1}{10}\{1732\tfrac{1}{2}p+810p\}$$
$$= \frac{2542\frac{1}{2}}{10}p \frown \textbf{£2·54 (nearest 1p).}$$

When the average of a set of measurements, which differ only by fairly small amounts from one another, is required, the following artifice will often save labour.

Ex. 9. The numbers of children in 14 classes in a small school are 31, 34, 32, 28, 29, 33, 30, 27, 31, 31, 28, 30, 32, 34. What is the average number of children in a class?

If we look at the classes we see that they do not differ much from 30, which we will *assume* is the average. We get the following differences:

Difference between actual number and 30	
+	−
I	2
4	I
2	0
3	3
I	2
I	0
2	
4	
18	8

We put in the zeros to remind ourselves of the relevant classes.

From the table, the total difference from 30 is $+18-8 = +10$.

Divide by the number of classes, i.e. 14.

$\therefore$ *Average* excess over 30, of each class, is $+\frac{10}{14} \simeq +0.71$.

$\therefore$ Average size of a class is approximately $30+0.71 = \mathbf{30.71.}$

A curious anomaly can arise in calculating averages at cricket. It is the convention that when a batsman is 'not out' at the end of an innings, he is credited with any runs made, but is not counted as having had an innings. Consider the following example.

Brown scored in 8 successive innings 5, 4, 6, 2 not out, 4 not out, 3, 5 not out, 4.

$$\text{Number of runs scored} = 5+4+6+2+4+3+5+4$$
$$= 33$$
$$\text{Number of completed innings} = 5$$
$$\therefore \text{Average} = \tfrac{33}{5} = \mathbf{6.6.}$$

His average is therefore more than the numbers of runs he actually made in any innings! The fault lies in the system adopted; that an incomplete innings should not be counted is a purely arbitrary rule, and is not based on mathematical principles. The result can hardly be expected to bear close analytical scrutiny.

EXERCISE 4

1. Find the average of £624, £561, £733, £247, £995.
2. 35 kg of pekoe at £4·80 per kilogramme are mixed with 210 kg of bohea at £2·24 per kilogramme. Find (a) the ratio of the cost of the pekoe to that of the bohea in the mixture, in simplest terms, (b) the proportion of pekoe in the mixture, in the form $x:1$, giving the result correct to two decimal places, (c) the value in pence per kilogramme correct to the nearest 1p, of the mixture of teas.

3. If it takes 42 minutes to cover a journey at an average speed of 48 km/h, how long will it take to do the journey at an average speed of 56 km/h?

4. A census is being made as to the number of passengers who catch buses in a particular area of a town. It is found that the average number on the 6 buses which serve on No. 1 route is 41, the average of the 8 buses which serve on No. 2 route is 37, and the average of the 3 buses which serve on No. 3 route is 25. What is the average number of passengers carried on a bus in this part of the town?

5. A train travels 30 km at an average speed of 64 km/h and a further 25 km at an average speed of 72 km/h. Find the average speed for the whole journey.

6. Smith's batting averages for 8 completed innings is 16·5. He bats for a ninth innings, also completed, and his new average for all 9 innings is found to be 18. How many did he score in his ninth innings?

7. The number of voters in the constituencies in a city are 22 571, 24 063, 23 492, 22 788, 23 361, 24 157 respectively. Find the average number of voters (to the nearest whole number).

  If 5 out of every 8 people voted during a general election, how many people did not vote in the city?

CHAPTER NINE

PERCENTAGE, PROFIT AND LOSS

1. Percentage

When we wish to *compare* quantities, fractions are not always helpful. Nor are decimals necessarily suitable for quick calculation. For example, suppose we wish to compare the following:

(a) A profit of $12\frac{1}{2}$p on a cost price of 75p
(b) A profit of 10p on a cost price of $62\frac{1}{2}$p
(c) A profit of 15p on a cost price of £1.

We need the fractions $\dfrac{12\frac{1}{2}}{75}$, $\dfrac{10}{62\frac{1}{2}}$, $\dfrac{15}{100}$.

We could bring them to the same denominator, or express each as decimals. Neither is convenient for some purposes. In such cases we arrange matters so that all fractions considered have a denominator of 100.

Now (a) $\dfrac{12\frac{1}{2}}{75} = \dfrac{25}{150} = \dfrac{25}{150} \times \dfrac{100}{100}$

$$= \dfrac{25}{\underset{3}{\cancel{150}}} \times \overset{2}{\cancel{100}} \ per\ 100$$

$= \frac{50}{3}\% = \mathbf{16\frac{2}{3}\%}$ (we write the denominator 100 as the symbol %, called 'per cent.' from the Latin *per centum* = out of 100).

Similarly (b) $\dfrac{10}{62\frac{1}{2}} = \dfrac{20}{125} \times 100\%$
$$= \mathbf{16\%}$$

and   (c) $\dfrac{15}{100} = \dfrac{15}{100} \times 100\%$
$$= \mathbf{15\%}.$$

∴ The best profit in the above three transactions is (a) relative to the others. It is the best return on outlay. It is not necessarily the largest cash profit; (c) is the largest.

We say that a profit of 15p on an outlay of £1 represents a 15% profit (fifteen per cent. profit).

The symbols C.P. for cost price, and S.P. for selling price will be used frequently. The difference between them is the profit made.

$$\text{i.e. S.P.}-\text{C.P.} = \text{Profit.}$$

Ex. 1. Smith obtains 17 marks out of a possible 40 in a test. What percentage is this?

$$\tfrac{17}{40} = \tfrac{17}{40}\times 100\% = 42\tfrac{1}{2}\%.$$

Ex. 2. Express $17\tfrac{1}{2}\%$ as: (a) a fraction in lowest terms; (b) a decimal.

$$(a)\ \ 17\tfrac{1}{2}\% = \frac{17\tfrac{1}{2}}{100} = \frac{35}{200} = \frac{7}{40}.$$

$$(b)\ \ \tfrac{7}{40} = 0{\cdot}175. \qquad\qquad 4)\underline{0{\cdot}700}$$
$$\phantom{(b)\ \ \tfrac{7}{40} = 0{\cdot}175. \qquad\qquad 4)}0{\cdot}175$$

The following percentages of £1 are very useful to remember:

$$5\% \text{ of } £1 = \tfrac{5}{100}\times £1 = \tfrac{5}{100}\times 100\text{p} = \textbf{5p.}$$

Similarly **10% of £1 is 10p in the £** and so on. This is one of the advantages of the new currency.

Ex. 3. Find the value of 6% of £364·62.

We require $\quad £\tfrac{6}{100}\times 364{\cdot}62$
$$= £6\times 3{\cdot}6462$$
$$= £21{\cdot}8772$$
$$\frown \textbf{£21·88} \text{ (to the nearest 1p).}$$

EXERCISE 1

1. Express the following as percentages:
   (a) $\tfrac{1}{5}$ (b) $\tfrac{1}{4}$ (c) $\tfrac{2}{3}$ (d) $\tfrac{7}{11}$ (e) $1\tfrac{1}{8}$ (f) $\tfrac{3}{700}$.

2. Express the following as fractions in lowest terms:
   (a) 15%   (b) 28%   (c) $27\tfrac{1}{2}\%$ (d) 160%   (e) $33\tfrac{1}{3}\%$ (f) 0·4%.

3. Express the following as decimals, correct to 4 decimal places, where not exact:
   (a) 35%   (b) $33\tfrac{1}{3}\%$ (c) $107\tfrac{1}{2}\%$ (d) $3\tfrac{3}{4}\%$ (e) $1\tfrac{7}{11}\%$ (f) $3\tfrac{3}{7}\%$.

4. Express (a) 14 as a percentage of 45
           (b) $12\tfrac{1}{2}$p as a percentage of £1
           (c) $27\tfrac{1}{2}$p as a percentage of £3·25
           (d) 245 g as a percentage of $3\tfrac{1}{2}$ kg.

5. Find the value of:
   (a) $7\tfrac{1}{2}\%$ of £120   (b) $2\tfrac{1}{4}\%$ of £85
   (c) 16% of $3\tfrac{1}{4}$ litres   (d) $8\tfrac{3}{4}\%$ of £90·48.

It is worth memorising the following relationships.

$100\% = 1 = 1$	$5\% = \frac{1}{20} = 0.05$
$50\% = \frac{1}{2} = 0.5$	$33\frac{1}{3}\% = \frac{1}{3} = 0.33\ldots$
$25\% = \frac{1}{4} = 0.25$	$66\frac{2}{3}\% = \frac{2}{3} = 0.66\ldots$
$10\% = \frac{1}{10} = 0.1$	$75\% = \frac{3}{4} = 0.75$

2. Percentage Profit

Suppose a soldering iron were bought for £2 and sold at a profit of 25%.

$$\text{The profit} = 25\% \text{ of Cost Price}$$
$$= \tfrac{25}{100} \times \text{C.P.}$$
$$= £2 \times \tfrac{25}{100} \text{ (for C.P.} = £2)$$
$$= £\tfrac{1}{2} = \mathbf{50p.}$$

Also the S.P. $=$ C.P. $+$ profit $= £2 + 50\text{p} = £2.50.$

We see at once that this is a particular case of the formula:

$$\mathbf{Profit = \frac{Percentage\ Profit}{100} \times C.P.} \qquad \circ \quad \circ \quad (1)$$

In another form, this gives on rearrangement:

Percentage Profit $\times$ C.P. $= 100 \times$ Profit

$$\text{i.e. } \mathbf{Percentage\ Profit = 100 \times \frac{Profit}{C.P.}} \qquad \circ \quad \circ \quad (2)$$

Ex. 4. An article is bought for £3 and sold for £3·87½. What is the percentage profit?

C.P. $= £3$, S.P. $= £3.87\frac{1}{2}$ ∴ Profit $= 87\frac{1}{2}\text{p}$

$$\therefore \text{Percentage profit} = \frac{87\frac{1}{2}\text{p}}{£3} \times 100\%$$

$$= \frac{87\frac{1}{2}\text{p}}{300\text{p}} \times 100\%$$

$$= \frac{175}{600} \times 100\%$$

$$= \mathbf{29\tfrac{1}{6}\%.}$$

Now it is often important to compare cost price and selling price of goods. Suppose that in a certain transaction a profit of $p\%$ is made (i.e. percentage profit is p). The *profit* is always worked on the *cost price, not* on the selling price.

If, then, the cost price had been 100, the selling price would have been 100+p. (For p% profit means p pounds profit for every 100 pounds outlay.)

Now formula (1) above states that

$$\text{Profit} = \frac{\text{Percentage Profit}}{100} \times \text{C.P.}$$

but Profit = S.P.—C.P., and Percentage Profit is p.

$$\therefore \text{S.P.—C.P.} = \frac{p}{100} \times \text{C.P.}$$

$$\therefore 100 \times \text{S.P.} - 100 \times \text{C.P.} = p \times \text{C.P.}$$

So $100 \times \text{S.P.} = (100+p) \times \text{C.P.}$

$$\therefore \textbf{S.P.} = \frac{\textbf{100}+\textbf{\textit{p}}}{\textbf{100}} \times \textbf{C.P.} \qquad \bullet \quad \bullet \quad (3)$$

$$\text{or } \textbf{C.P.} = \frac{\textbf{100}}{\textbf{100}+\textbf{\textit{p}}} \times \textbf{S.P.} \qquad \bullet \quad \bullet \quad (4)$$

These formulæ are very useful and are easily understood and remembered if the work is laid out as under.

Ex. 5. A purse is sold for £3·60 at a profit of 8%. How much did it cost originally?

If the C.P. had been 100, the S.P. would have been 108.

Let the actual C.P. be £x; the actual S.P. is £3·60.

C.P.	S.P.
£x	£3·60
100	108

By direct comparison

$$\frac{x}{100} = \frac{£3·60}{108}$$

$$\therefore x = \frac{£100 \times 3·60}{108} \text{ (as in formula (4) above)}$$

$$= £\frac{360}{108} = £\frac{10}{3}$$

$$\backsimeq \textbf{£3·33} \text{ (to the nearest 1p).}$$

Ex. 6. As a result of evacuation during hostilities the population of a town decreased to 79 200, representing a drop of 12%. What was the previous population?

F.P.	S.P.
x 100	$\dfrac{79\ 200}{88}$

F.P. = First Population
S.P. = Second Population

As before, let F.P. $= x$

$$\therefore \frac{x}{100} = \frac{79\ 200}{88}$$

$$\therefore\quad x = \frac{100 \times 79\ 200}{88}$$

$$= 90\ 000.$$

Ex. 7. A toy costing 45p is sold at a profit of 20%. What is the selling price?

C.P.	S.P.
45p 100	x 120

Here the S.P. ($=x$, say) is the unknown quantity.

$$\therefore \frac{x}{120} = \frac{45p}{100}$$

$$\therefore\quad x = \frac{120 \times 45}{100}p$$

$$= 6 \times 9p = \textbf{54p.}$$

Ex. 8. An author receives £280 on account for a set of articles. If this represents 35% of the amount due altogether, how much more can he expect to receive?

£280 is 35% of the total, so that 65% has still to be paid (to make the total of 100%).

Let £x be the remainder due.

Amount Paid	Amount Due
28 35	x 65

$$\therefore \frac{x}{65} = \frac{280}{35}$$

$$\therefore x = \frac{\overset{40}{\cancel{280}} \times \overset{13}{\cancel{65}}}{\underset{\underset{1}{5}}{\cancel{35}}} = 52.$$

∴ **There is £520 more payment due later.**

EXERCISE 2

1. Find the selling price if:
   (a) the cost price is £4, profit is 20%
   (b) the cost price is £3·75, profit is 16%
   (c) the cost price is 62½p, loss is 12%.

2. Find the percentage profit of loss if:
   (a) cost price is £4, selling price is £4·50
   (b) cost price is £4, selling price is £3
   (c) cost price is £2·08, selling price is £2·73.

3. Find the cost price if an article is sold for:
   (a) £3 at a profit of 20%
   (b) £3·20 at a loss of 20%
   (c) £1·32 at a profit of 33⅓%.

4. Find the value of the following, correct to the nearest ½p:
   (a) 2½% of £7·82
   (b) 5½% of £34·57
   (c) 9¼% of £504·91.

5. The diameter of the earth is 7913 miles and that of the sun is 864 367 miles. What percentage of the sun's diameter is that of the earth? (Correct to 2 significant figures.)

6. During a year Smith spends £4104 which represents 95% of his income. How much does he save?

7. The rate in the pound charged by a local authority is found to be 68p. Of this, household refuse removal costs 2·64p. What percentage of the rate in the pound is this? (See Ch. 15.)

8. To insure a building under a comprehensive policy, the premium asked by a particular insurance company is 23p per £100 insured. What percentage is this of the sum insured? What would it cost to insure a house worth £21 250?

3. Harder Percentage Problems

There are extensions of the principles indicated above which have practical applications in everyday life.

We shall first consider the changes which take place in successive transactions. Goods are sold by a manufacturer to a retailer, who in turn sells them to a customer. There is a profit made on each part of the business, but in dealing with the percentage change in price *we may not add the successive percentages together*. They lead to *multiplication* as will be seen from the following example.

Ex. 9. Smith & Co. make stools at a cost of £3·50. They sell them to Brown & Tomkins' Furniture Stores at a profit of 20%. The price at which they are sold to a customer ensures a profit of 25% for Brown & Tomkins. What is the retail price and how much per cent. is this greater than the original cost price?

C.P. Manufacturer's cost
S.P.1 Wholesale selling price
S.P.2 Retail selling price

C.P.	S.P.1	S.P.2
£3·50		
100	120	
	100	125

We see at once that

$$\frac{\text{S.P.2}}{\text{S.P.1}} = \tfrac{125}{100}, \text{i.e. S.P.2} = \tfrac{125}{100} \times \text{S.P.1}$$

Also $\dfrac{\text{S.P.1}}{\text{C.P.}} = \tfrac{120}{100}, \text{i.e. S.P.1} = \tfrac{120}{100} \times \text{C.P.}$

$$\therefore \text{S.P.2} = \tfrac{125}{100} \times \tfrac{120}{100} \times \text{C.P.}$$
$$= \tfrac{125}{100} \times \tfrac{120}{100} \times £3·50$$
$$= £5·25.$$

Also $\dfrac{\text{S.P.2}}{\text{C.P.}} = \tfrac{125}{100} \times \tfrac{120}{100} = \tfrac{3}{2}.$

∴ If C.P. is 100, S.P.2 is $\tfrac{3}{2} \times 100 = 150.$

∴ Percentage increase is **50%**. Not 45% as would have been obtained by adding the successive increases per cent. We could, if we had known them, have added the *actual* profits, and found the percentage increase altogether as

$$\frac{\text{1st Profit} + \text{2nd Profit}}{\text{Cost Price}} \times 100.$$

Let us next investigate the percentage error made in measuring lengths, areas and volumes. Now error is the difference between calculated and true values, so percentage error is given by

Percentage error

$$= \frac{\text{Calculated value} \sim \text{True value}}{\text{True value}} \times 100 \quad \bullet \quad \bullet \quad (1)$$

The symbol $\sim$ means difference between the quantities on either side of it, putting the larger one first. It is not difficult to see that in practice the true value may not be known exactly, although the greatest margin of error may be. In this case a practical modification of (1) would be

Percentage error *is less then*

$$\frac{\text{Maximum Error}}{\text{Calculated Value}} \times 100 \quad \bullet \quad \bullet \quad \bullet \quad (2)$$

The relationship (2), which is not an equation, can lead to some fairly difficult problems, so we shall content ourselves with applications of equation (1) and very simple cases of (2). It is instructive to think for a moment of an application of (2), however, as in measurement with a ruler. A ruler is not an *exact* instrument. Suppose we find that the length of a line is 8 in. We can assume that the ruler is accurate within $\frac{1}{10}$ of an inch. Therefore, the percentage error in taking the length of the line as a true 8 in is likely to be less than

$$\frac{0 \cdot 1}{8} \times 100\%, \text{ i.e. } 1\tfrac{1}{4}\%.$$

We cannot measure 8 in exactly, however hard we try, so we cannot find the exact percentage error.

Ex. 10. A rectangular field 80 m long and 120 m wide is called by Farmer Brown a hectare field. What is the percentage error?

Actual area $= 80 \times 120 = 9600 \text{ m}^2$
Area claimed $= 1$ hectare $= 10\ 000 \text{ m}^2$

$$\therefore \text{Percentage error} = \frac{10\ 000 - 9600}{9600} \times 100$$

$$= \frac{400 \times 100}{9600} = 4\tfrac{1}{6}\%$$

i.e. the field is $4\tfrac{1}{6}\%$ smaller than claimed.

Discount. Discount is a reduction given on the normal price. A shopkeeper may offer a reduction in the marked price of goods during a sale. A typical example would read 'During this week there is a reduction of 10% on all marked prices'. It could equally well read '10% discount on all marked prices during this week'.

Now 10% of £1 = 10p; and the sale price is 100—10 = 90 per cent. of the marked price, i.e. 90p has to be paid out of each £1 marked.

Alternatively, we have

$$\text{Sale Price} = \tfrac{90}{100} \times \text{Retail Price}.$$

Ex. 11. A tape recorder is sold for £60·90 normally, but a discount of 5% is allowed for immediate cash payment. How much would a customer pay if he offered outright purchase?

Normal charge = £60·90.

Reduction of 5% means that the reduced charge is $\tfrac{95}{100}$ of the normal charge.

$$\therefore \text{Cash payment} = £60\cdot90 \times \tfrac{95}{100}$$
$$= £60\cdot90 \times 0\cdot95$$
$$\simeq £57\cdot85.$$

$$
\begin{array}{r}
60\cdot9 \\
0\cdot95 \times \\
\hline
3\cdot045 \\
54\cdot81 \\
\hline
57\cdot855
\end{array}
$$

(The reader will note that, as so often happens, we use more decimal places in money than we require in the answer. The interpretation of £0·005 as 0p or 1p seems to be something of a matter of courtesy.)

EXERCISE 3

1. During each year a car loses 15% of its value at the beginning of the year. What is its value in January 1979, if it cost £2400 in January 1976?

2. Smith has £1000 worth of goods. He sells £850 worth at 10% profit and the remainder at 5% loss. What percentage profit did he make on the whole transaction?

3. A shopkeeper sells a clock for £30, thereby making a profit of 20%. During a sale he allows a discount of 10% on the marked price. What is his new percentage profit?

4. A merchant pays £13·12½ for a pair of steps. He normally sells them at a profit of 20%, but allows a discount of 5p in the pound during a January sale. What is the sale price (to the nearest penny)?

5. A room measured roughly as 10·5 m by 12·5 m. Its true area is 120 m². What is the percentage error in measurement of the area?

6. A grocer buys sugar at £9·70 for a 50 kilogramme chest. He sells the sugar at 28p a 1 kg bag, but 5% is wasted on making it up into bags. What is (a) his actual profit, (b) his percentage profit on one chest?

7. What percentage increase is necessary to change a loss of 5% into a profit of 5%?

8. A household spends one-sixth of his income on rent, one-quarter on food and one-ninth on clothes. He spends £3116 on these items in 1977. During 1978 his rent increases by 5%, his food costs by 20%, and the cost of clothes by 17%. How much does he spend on these in 1978? What is the overall percentage increase on rent, food and clothes?

4. Mixtures

An introduction to problems involving mixtures was given in the previous chapter, under the section on averages. We will now consider inverse problems on mixtures, and include percentage profit.

Ex. 12. In what proportion should coffee at £6·60 a kg be blended with coffee at £7·40 a kg, if the mixture is to be sold at £8·50 a kg, thereby making a profit of 25%?

We have $\dfrac{\text{C.P. of mixture}}{\text{S.P.}} = \dfrac{100}{125} = \dfrac{4}{5}$

$\therefore$ C.P. of mixture $= \dfrac{4}{5} \times 850\text{p}$
$= 680\text{p}$.

Coffee at £6·60 a kg costs 20p *less* than mixture.
Coffee at £7·40 a kg costs 60p *more* than mixture.
$\therefore$ **3 parts of the cheaper coffee are mixed with 1 part of dearer.**
[For in each 4 kg, $3 \times 20\text{p}$ ($= 60\text{p}$) too little is charged and $1 \times 60\text{p}$ ($= 60\text{p}$) too much is charged.]

Alternatively, the second part of the calculation can be done very neatly as follows:

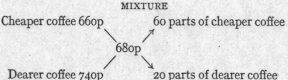

MIXTURE

Cheaper coffee 660p   60 parts of cheaper coffee

680p

Dearer coffee 740p   20 parts of dearer coffee

i.e. number of parts of cheaper coffee to number of parts of dearer coffee is 60:20, i.e. 3:1.

EXERCISE 4

1. Coffee at $81\frac{1}{2}$p per 100 grammes is mixed with coffee at $75\frac{1}{2}$p per 100 grammes. In what proportion should they be mixed, if the resultant blend is worth 78p per 100 grammes?

2. 15 kg of tea at £2·52 a kg are mixed with 21 kg of tea at £2·80 a kg. At what price per kilogramme must the mixture be sold to make a profit of 20%?

3. Find the percentage profit on selling, at 10½p per kilogramme, a commodity which costs £4·35 for a 50 kg chest of it.

4. If milk costing 27p a litre is mixed with water, of negligible cost, in the ratio of 5:1, and the mixture is sold at 30p a litre, what percentage profit is made? (Not a very reputable dairy, this!)

5. Wine at £1·62 a half litre is mixed with wine at £1·16 a half litre. In what proportion are they mixed, if the mixture is sold for £2·40 a half litre, giving a profit of 25%?

6. In a club of 80 members, 62 play bridge, 74 attend the club dances and 55 play tennis on the club courts. What is the *least* possible number of members who take part in all three activities?

7. Fifty-three passengers are riding on a bus. Some take 15p tickets and the rest have 25p tickets. Altogether their fares come to £10·15. How many 25p tickets were issued?

CHAPTER TEN

ARITHMETICAL GRAPHS

1. Plotting and Interpreting Graphs

Information provided in words often takes some time to assimilate. If, however, we can represent the situation pictorially we can grasp quickly the salient features present. Reading a set of numbers gives no clear image to remember, whereas a diagram will be retained in mind for a considerable time.

It is accepted practice nowadays to give diagrams and pictures wherever possible to clarify important points in many forms of study, particularly when scientific or technical. For example, only a limited number of people would read a book on animal life, and having read it still less would not have forgotten most of the contents within a very short space of time. On the other hand, how vast a number is enthralled by natural-history films! It requires no great mental effort to sit back in a cinema, watching a film: reading is a sterner test of self-discipline. Visual aids form an integral part of modern education and graphs are a form of visual aid.

We will assume that we are given two quantities x and y, where y changes in value when x changes. If we wish to mark a point on a diagram to represent a *particular* pair of values of x and y, we can do so by measuring the distance of the point from each of two perpendicular straight lines, called the *axes*. We measure x horizontally (in the direction OX in fig. 1) and y vertically (in the direction OY in fig. 1). The point O where the axes meet is called the *origin*. The lines OX, OY are the horizontal and vertical axes respectively.

Suppose a cyclist is travelling at 20 km/h. He will go 20 km in 1 hour, 40 km in 2 hours and so on. We can plot a series of points, laying off time (in hours) horizontally, and distance (in kilometres) vertically. Our first point P is found by measuring 1 hour horizontally and then laying off 20 km vertically. We call the point P (1, 20); the numbers 1 and 20 are the *coordinates* of the point. Q is the point (2, 40) and is given by 2 hours horizontally and 40 km vertically. Similarly we get R (3, 60) etc.

If we then join P, Q, R . . . we find that we have a straight line. Further, this line passes through O, as is reasonable, for the cyclist will have gone 0 miles in 0 hours. If we wish to know how far he would go in $2\frac{3}{4}$ hours we lay off a vertical line HK, through H, the

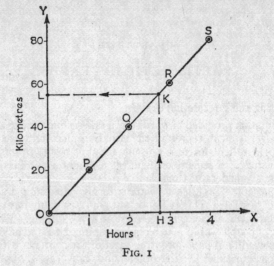

FIG. I

point $2\frac{3}{4}$ on OX, meeting the straight line graph at K. We then draw KL horizontally, meeting OY at L. OL (= 55 km) is the required distance.

This is a very simple example of an arithmetical graph. Graphs take various forms. Some consist of separate parallel lines (bar charts), others are polygonal (that is, they consist of connected lines joining a set of points in order. If A, B, C, D . . . is the set of points in order, then the lines AB, BC, CD . . . form a polygonal graph). Yet again, other graphs are smooth curves. The type of graph used depends on the nature of the information provided (see fig. 2).

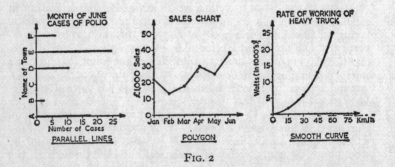

FIG. 2

It is worth bearing in mind two points when drawing graphs:

(1) It is not essential for the origin O to be the point where both quantities being laid off shall be zero. Sometimes it is in fact undesirable for it to be zero.

(2) The scales for x and y need carefully choosing so as to make the graph as large and well shaped as possible, bearing in mind that the necessary points must all appear on the graph paper. Fig. 3 illustrates these two points.

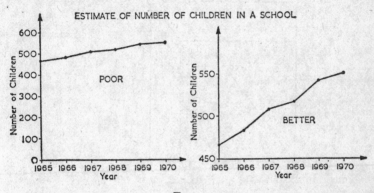

FIG. 3

Ex. 1. A motorist travels an hour and a half at an average speed of 50 km/h and then he stops for half an hour. He then continues his journey at a speed of 50 km/h. Find how far he has gone $2\frac{1}{4}$ hours after starting.

If he started out at 10.0 a.m. and a motor-cyclist starts on the same journey at 11.30 a.m. and travels at 90 km/h, find when and where the motor-cyclist overtakes the motorist.

We use graph paper marked out in large squares of 1 unit and small squares of $\frac{1}{10}$ unit side. Mark off hours along OX, taking a scale of 1 unit to represent 1 h. Lay off distance in miles along OY, taking a scale of 1 unit to represent 40 km.

We see that the motorist's journey consists of three straight-line parts and the motor-cyclist's graph is a single straight line beginning $1\frac{1}{2}$ hours later. We plot the points (0, 0) and (1, 50) for the motorist, stopping this line $1\frac{1}{2}$ hours after starting: we then lay off a horizontal line for half an hour as he has not changed his distance: finally we lay off a line through (2, 75) and (3, 125) representing the last stage in his journey. The points $(1\frac{1}{2}, 0)$, $(2\frac{1}{2}, 90)$ will give the motor-cyclist's graph.

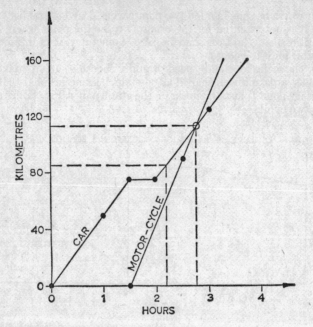

We see from graph that (1) the motorist is 87½ km on his way after 2¼ hours (2) he is overtaken by the motor-cyclist at 12.45 p.m. at a distance of 112½ km from their starting point.

Ex. 2. A patient with a high fever has his temperature taken every four hours. From the following set of results draw a temperature graph. Find from it the highest temperature. Does the fever show signs of abating or not?

Time	Noon	4 17.11.70	8	Mid-night	4	8	Noon 18.11.70	4.	8	Mid-night
Temp °C	38·3	38·6	38·8	39·1	39·2	38·8	38·6	38·7	39·2	39·4
Time	4	8	Noon 19.11.70	4	8	Mid-night	4	8	Noon 20.11.70	4
Temp. °C	39·4	39·0	38·5	38·4	38·6	38·7	38·4	37·9	37·9	37·9

We lay off time horizontally using 1 cm for 4 hours: temperature vertically, using 2 cm for 1°C.

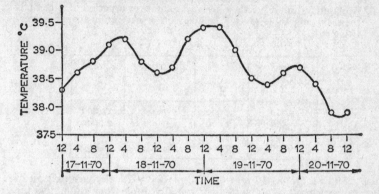

From the graph we see that the highest temperature is 39·45° C and that it occurs at about 2 a.m. on 19th November 1970. The highest temperature on the following day was only 38·7° C and the fall appears to be maintained. The graph suggests that the fever has passed its peak.

EXERCISE 1

1. A stone is dropped down a well and the distance fallen at various times is given by the following table.

Time taken (sec.)	0	0·5	1·0	1·5	2·0	2·5
Distance fallen (m)	0	1.23	4·91	11·04	19·62	30·69

If the splash, when the stone strikes the water, is heard 1·8 seconds after releasing the stone, estimate the depth of the water level below ground, from a graph. If the well had been 24 m deep to water level, how long would it have taken before the splash was heard? (Neglecting time for sound to travel.)

2. It is 291 km from London to Manchester. An express train leaves London at 8.40 a.m. for Manchester and travels at 83 km/h. A goods train leaves Manchester for London at 10.10 a.m. and travels at 55 km/h. Find when they will pass one another, and their distance from London at that moment, by drawing a graph. (The graph consists of two straight lines intersecting.)

3. A manufacturer advertises his wares and over a period of time finds the following table gives a comparison of net profit against advertising cost.

Advertising Cost £	Profit £	Advertising Cost £	Profit £
2000	4 000	7 400	22 000
3500	7 000	9 600	22 000
5800	13 000	11 400	20 800
6800	20 000	14 000	18 000

Draw a graph of Profit (in £'s), plotted vertically, against cost of Advertising (in £'s), plotted horizontally. From your graph, find the best amount to pay in advertising.

How do you account for the shape of the graph?

2. Histograms and Frequency Distributions

Sometimes the data are of such a nature that we do not know precise points to plot. This often occurs in statistical work. In the measurement of heights of children of a particular age, it might be found that 4 were between 1·57 m and 1·58 m, 12 were between 1·58 m and 1·59 m and so on. It would not be practicable to indicate every precise height, nor would it be of any real value. In such a case we plot frequency of observation against the observation range, and build up a diagram known as a histogram.

We will illustrate the principle by an example.

The following table indicates the marks obtained out of 100 by 60 candidates for an examination.

TABLE A.

Maximum 100

10	33	43	82	39	14	70	60	62	64	54	29
47	22	41	77	57	62	36	71	54	59	81	53
53	17	58	45	72	55	43	37	51	50	3	47
64	59	74	66	64	51	49	40	25	38	60	36
58	60	91	68	83	60	58	22	48	77	72	58

As it is presented (the order in which the marks appeared on the papers when examined) the information conveys nothing to the reader. It is necessary to reorganise the marks in a systematic form, and to group them into suitable *classes*. For example, the first class might be the number of candidates who obtained from 0 to 9 marks, the second class the number who obtained between 10 and 19 inclusive, and so on. We then have:

TABLE B.

Inclusive mark range	Frequency	Inclusive mark range	Frequency
0– 9	1	50–59	15
10–19	3	60–69	11
20–29	4	70–79	7
30–39	6	80–89	3
40–49	9	90–99	1

We now construct a histogram of frequency against mark range.

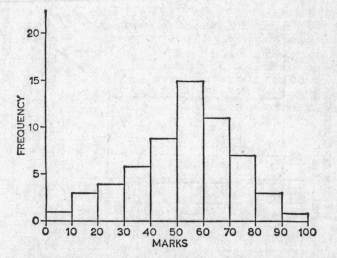

This diagram clearly indicates the distribution of marks obtained. The marks could, however, have been grouped under an entirely different schematic arrangement.

From the previous table B construct a new table C by finding the number of candidates who obtained fewer than a specified number of marks.

TABLE C.

No. of candidates getting less *than*	Frequency	No. of candidates getting less *than*	Frequency
10	1	60	38
20	4	70	49
30	8	80	56
40	14	90	59
50	23	100	60

The table is constructed as follows (from table B):

$$
\begin{aligned}
1 &= 1 \\
1+3 &= 4 \\
1+3+4 &= 8 \\
1+3+4+6 &= 14, \text{ etc.}
\end{aligned}
$$

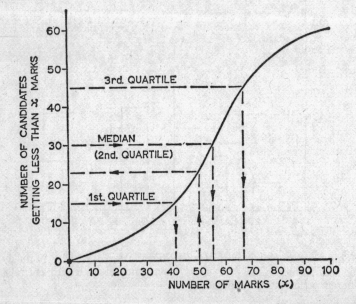

We then draw a smooth curve, called a *cumulative frequency curve* or *ogive*, with the number of candidates getting fewer than x marks plotted vertically against x marks plotted horizontally (Graph p. 94).

From the ogive, various interesting results can be read off at once. The *median mark* is the mark obtained by the $\frac{60}{2} = $ 30th candidate, and is 55, so half of the candidates obtained over 55%. (Strictly speaking, if there are n candidates, the median mark is that of the $\frac{n+1}{2}$th candidate).

The *quartiles* divide the distribution into 4 equal parts. There are three quartiles, of which the middle one is the median already defined. We can alternatively calculate the median mark from the table C: 23 candidates obtained less than 50 marks, 38 obtained less than 60 marks, therefore a good estimate of the mark of the 30th candidate is

$$50 + \left(\frac{30-23}{38-23}\right) \times (60-50) = 50 + \tfrac{7}{15} \times 10$$

$$= 54 \cdot 7 \backsimeq \mathbf{55}, \text{ which agrees}$$

with the graphical result.

Suppose the pass mark to be 50%. We see from the ogive that the number of candidates who failed is 23, so $60 - 23 = 37$ passed.

∴ The percentage of success was

$$\tfrac{37}{60} \times 100 = \mathbf{61 \cdot 7\%}.$$

Exercise 2

1. In a certain district it is found that the numbers of children in families gives a frequency distribution as follows:

Number of Children in the Family	0	1	2	3	4	5	6	7	8
Frequency[11]	82	147	251	107	40	9	3	1	0

Draw (*a*) a histogram, (*b*) an ogive. From the latter find the median, and indicate its position in the former.

2. Take table A (on page 92) and from it construct a frequency table like table B, but using mark ranges 0–4, 5–9, 10–14 etc., and from your table plot a histogram. Compare it with that given on page 93. Comment on their difference in appearance.

[11] i.e. Number of families in which this occurs.

3. Construct an ogive for the following table of heights, and find the *median* height.

Height in Metres[12]	Frequency
1·55	1
1·60	4
1·65	12
1·70	22
1·75	17
1·80	9
1·85	5
1·90	1

Compare the result with the *mean* height, which is the same as the average. It is found as follows:

$$\text{Mean} = \frac{\text{Total of all the heights}}{\text{Number of heights}} = \frac{f_1 \times h_1 + f_2 \times h_2 + \ldots}{f_1 + f_2 + \ldots}$$

where f_1 is first frequency, h_1 is first height; etc.

[12] The height given is the mid point of the interval; e.g. 1·75 m includes people from 1·725 m to 1·775 m in height.

CHAPTER ELEVEN

MENSURATION OF RECTANGULAR FIGURES

1. The Area of a Rectangle

Mensuration is the measurement of geometrical quantities, such as lengths of lines, areas of plane or curved surfaces, and volumes of solids.

In this chapter we shall consider *rectangular* plane figures and solids, and applications of these shapes to problems.

A rectangle is a plane figure, popularly known as an oblong, having the properties that its opposite sides are equal in length and parallel to one another, and all its angles are right angles.[13] If all the sides of the rectangle are equal, then the figure is a square.

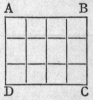

If we have a square, 1 cm long and 1 cm broad, we say that its area is 1 *square* centimetre (cm²).

Suppose now that we have a rectangle ABCD, 4 cm long and 3 cm broad. Draw lines 1 cm apart dividing it into 1-cm squares as shown. Each row contains 4 squares, but there are 3 rows.

∴ The number of squares = $4 \times 3 = 12$ squares, and we say that the area is 12 cm².

In general if the length of a rectangle is l cm and its breadth is b cm, its *area* $= l \times b$ cm² . . . . (1)

In formula (1) if the area is A cm² we have

$$A = l \times b \text{ (more shortly, } A = lb)$$
$$\therefore l \times b = A$$

$$\text{so } l = \frac{A}{b} \text{ and } b = \frac{A}{l}$$

on dividing both sides by either b or by l in turn.

In words: Length $= \dfrac{\text{Area of Rectangle}}{\text{Breadth}}$ . . . . (2)

$\phantom{\text{In words: }}$Breadth $= \dfrac{\text{Area of Rectangle}}{\text{Length}}$ . . . . (3)

The formulæ (1), (2) and (3) are all the same really, but the *subject*

[13] A *minimum* definition of a rectangle is much more restricted than this, namely, that it is a parallelogram having one right angle.

of the formula (i.e. the unknown quantity to be found) has been changed.

2. Perimeter

The *perimeter* of *any* figure is the length round the edge.

In rectangle PQRS, the perimeter is

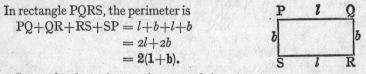

$$PQ+QR+RS+SP = l+b+l+b$$
$$= 2l+2b$$
$$= 2(l+b).$$

In words, *the perimeter of a rectangle* is twice the sum of its length and breadth.

The perimeter of a *circle* is the length of its circumference, and is $2\pi r$ (where r is the radius), as we shall see later. The perimeter of a *polygon* (a figure bounded by straight lines) is the sum of the lengths of its sides.

Ex. 1. Find the perimeter and area of a rectangle 6 m long and 5 m wide.

$$\text{Area} = \text{Length} \times \text{Breadth}$$
$$= 6 \times 5 \text{ m}^2$$
$$= \mathbf{30m^2}$$
$$\text{Perimeter} = 2 \text{ (Length+Breadth)}$$
$$= 2 \text{ (6+5) m}$$
$$= \mathbf{22 \ m.}$$

Ex. 2. Find the area of a rectangle 2·64 m by 3·45 m.

$l = 3{\cdot}45$ m, $b = 2{\cdot}64$ m
$\therefore A = lb = 3{\cdot}45 \times 2{\cdot}64$ m^2
$\qquad \frown \mathbf{9{\cdot}11 \ m^2}$ (to 2 decimal places).

$$
\begin{array}{r}
2{\cdot}64 \\
3{\cdot}45 \times \\
\hline
1320 \\
1{\cdot}056 \\
7{\cdot}92 \\
\hline
9{\cdot}108
\end{array}
$$

Note. It is worth remembering that on the Continent, where a century of experience has been gained, measurements of length are quoted in the *largest suitable unit in general use*, e.g.

(a) 7·4 cm, *not* 74 mm,
(b) 1·76 m for the height of a person, *not* 176 cm.

Corresponding procedure is adopted for area and for volume, e.g.

(c) 7460 cm^2 would be correct, but
(d) 7·46 m^2 would be used for 74 600 cm^2 (for 10 000 cm$^2 = 1$ m^2); one could, of course, use in 74 600 cm^2 the actual working of a problem.

Find the areas of the rectangles with the following dimensions (questions 1–5), giving the answer correct to 2 decimal places where necessary:

1. 7 cm long, 10 cm wide.

2. 3 m long, 2·4 m wide.

3. 1·62 m long, 4·77 m wide.

4. 12·85 km long, 36·4 m wide (answer in km²).

5. 6·42 m long, 7 mm wide (answer in cm²).

6. Find the perimeters in questions 1–5 above.

7. The area of a rectangle is 24 m². If it is 8 m long, what is its perimeter?

8. A rectangular room is of perimeter 24·2 m and length 7.5 m. What is its area?

9. A field is of area 9 hectares and is of rectangular shape. The length of one side is 360 m. Find the length of the perimeter.

10. A rectangular flower-bed 10 m by 8 m is increased to 10·5 m by 8·5 m. What is the percentage increase in area?

3. Applications to Plane Figures

If we wish to find the cross-sectional area of a body, such as the casting illustrated, it is easily found by dividing the area up into rectangles, as indicated by dotted lines.

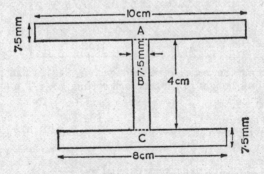

We have 7·5 mm = $\frac{3}{4}$ cm.

∴ Area required = A+B+C

$$= 10 \times \tfrac{3}{4} + 4 \times \tfrac{3}{4} + 8 \times \tfrac{3}{4} \text{ cm}^2$$

$$= (10+4+8) \times \tfrac{3}{4} \text{ cm}^2$$

$$= \frac{22 \times 3}{4} \text{ cm}^2$$

$$= 16\tfrac{1}{2} \text{ cm}^2.$$

When we have a piece removed from a lamina (i.e. a plane body like a piece of plywood, of almost negligible thickness), or if we wish to calculate, say, the area of a lawn surrounding a flower-bed, we subtract the removed (or inside) area from the whole area. We do not divide the remaining area into small rectangles, as this leads to unnecessary labour.

Ex. 3. A path of width $1 \cdot 5$ m is to be laid round a rectangular lawn 25 m by 14 m. Calculate the area of the path. (See diagram below.)

Length of whole region $= 25 + 1\frac{1}{2} \times 2 = 28$ m
Breadth of whole region $= 14 + 1\frac{1}{2} \times 2 = 17$ m

$\therefore$ Total area (path+lawn) $= 28 \times 17 = 476$ m^2
$\qquad$ Area of lawn $= 25 \times 14 = 350$ m^2

$\therefore$ Required area of the path $\qquad = \textbf{126 m}^2.$

Note that we add twice the width of the path in assessing the whole length and whole breadth.

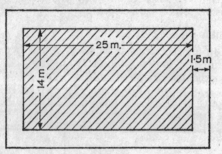

Ex. 4. Find the cost of fertilising a rectangular lawn 49 m by 21 m with fertiliser bought in packets each costing £2·52, if a packet will cover 90 m^2.

Area of lawn 49×21 m^2
Area covered by packet $= 90$ m^2

$$\therefore \text{No. of packets required} = \frac{49 \times \overset{7}{21}}{\underset{30}{90}}$$

$$= \tfrac{343}{30} = 11\tfrac{13}{30}.$$

$\therefore$ 12 *packets are required*, as shopkeepers are loath to sell $\frac{13}{30}$ of a packet.
$\qquad \therefore$ Total cost $= 12 \times £2 \cdot 52 = \textbf{£30·24.}$

A particularly useful application of the area of a rectangle is the assessment of the quantity of wallpaper required to paper a room, or the amount of distemper needed for a ceiling.

A '*piece*' of wallpaper is 10 m long and 50 cm wide.[14] Its area is therefore $10 \times 0.5 \, m^2 = \mathbf{5 \, m^2}$. The room to be decorated should be thought of as having the walls laid out flat. (Length $= l$ m, breadth $= b$ m, height $= h$ m.)

$$\text{The total length} = \text{perimeter of room}$$
$$= 2\,(l+b) \text{ metres}$$
$$\therefore \text{Area of walls} = \text{Total length} \times \text{height}$$
$$= \text{Perimeter} \times \text{height}$$
$$= \mathbf{2\,(l+b)\ h\ square\ metres.}$$

Although it is not customary to paper doors or windows, there is no necessity to deduct their areas from the quantity of paper estimated, unless they are very large. The difference is useful in making up for wastage. Another practical point is to remember to allow extra for matching patterned paper from one strip to the next!

EXERCISE 2

1. Find the areas of the given figures. All the angles are right-angles.

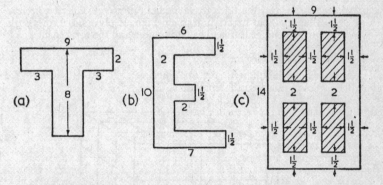

All the measurements in this question are in centimetres. The shaded areas have been cut out.

2. A photograph 25 cm by 22·5 cm is placed inside a frame of overall size 30 cm by 25 cm. What is the uncovered area surrounding the photograph? What is the ratio of this area to that of the photograph?

[14] This is not exact. At the time of writing, the author measured various wall-papers and found them to vary in width from 51·4 cm to 51·8 cm, after trimming. The 5m² is a safe guide and allows a small margin.

3. A lawn is 11 m by 8·4 m. Two rectangular flowerbeds are cut in it, each of size 3 m by 2·5 m. What is the area of the lawn remaining? How many turves 1 m by 30 cm would be needed to cover this remaining area of lawn, and what would be their cost at 3p each?

4. A rectangular room is 5·35 m long and 4·4 m wide. Its height to the picture rail is 2·3 m. Calculate the area of the four walls and find the number of pieces of wallpaper required to cover the walls.

If, now, it were decided to paper the end walls with paper of one kind and the side walls with paper of another sort, would it be necessary to order more paper? How many pieces of each sort would be needed?

5. The diagram represents a rectangular box. If all the external faces are to be painted except the base, what is the area to be dealt with? (Answer in square metres.)

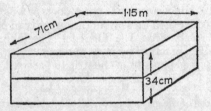

6. A packing case with a lid is to be made of plywood. The case is to be 1·05 m long, 68 cm broad and 57 cm high. If plywood costs 34½p a square metre, find the area of wood required and the total cost.

4. Volume of a Cuboid

A *cuboid* is a solid with rectangular faces. It can, if desired, be called a rectangular parallelepiped!

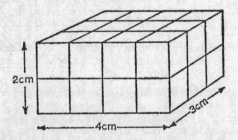

Consider a box 4 cm long, 3 cm wide and 2 cm high. It can be imagined as divided up into a number of cubes each 1 cm × 1 cm × 1 cm, i.e. each of volume 1 cm³.

The top layer has 4×3 of these cubes, and the second layer has the same. $\therefore$ Altogether there are $4 \times 3 \times 2$ cubes, i.e. **24 cubes,** but the volume of each cube is 1 cu. cm. $\therefore$ Volume of cuboid = **24 cm³**.

This result can clearly be extended to a cuboid of length l m, breadth b m and height h m, giving

$$\text{Volume of cuboid} = \mathbf{l \times b \times h\,m^3}$$
$$= (l \times b) \times h\,\text{m}^3$$
$$= \textbf{Area of base} \times \textbf{height}$$

If we call the volume, V m³, we have

$$l \times b \times h = V$$

$$\therefore h = \frac{V}{l \times b} \text{ (or more shortly } \frac{V}{lb})$$

$$\text{i.e. } \mathbf{h = \frac{V}{A}} \text{ (where } A \text{ is the base area).}$$

(This last formula, $h = \dfrac{V}{A}$ applies to many solids which are not cuboids. It does in fact apply to prisms in general. Thus, the formula $V = A \times h$ is of widespread use—e.g. in calculating the volume of material in a girder of given cross-sectional area and length.)

Ex. 5. The volume of a tank is 30 m³. Its length and breadth are 3·5 m and 2·4 m respectively. Find its height.

$$\text{Height} = \frac{\text{Volume}}{\text{Length} \times \text{Breadth}} = \frac{30}{3 \cdot 5 \times 2 \cdot 4}\,\text{m}$$

$$= \frac{30 \times 100}{35 \times 24}\,\text{m} = \frac{25}{7}\,\text{m} \simeq \textbf{3·57 m.}$$

Ex. 6. How many bricks 24 cm $\times$ 12 cm $\times$ 8 cm are required to build a wall 5·5 m long, 3·6 m high, and 24 cm thick (ignoring thickness of mortar)?

$$\text{Volume of brick} = 24 \times 12 \times 8\,\text{cm}^3$$
$$\text{Volume of wall} = 550 \times 360 \times 24\,\text{cm}^3$$

$$\therefore \text{Number of bricks} = \frac{550 \times 360 \times 24}{24 \times 12 \times 8}$$

$$= 2062\tfrac{1}{2}$$

i.e. 2063 bricks are required.

Notes. (1) The cancellation of the 24 cm length is not artificial for the length of a brick is equal to the thickness of a double-width wall.

(2) Simplification is delayed until the end.

(3) The actual number of bricks required in practice would have to be greater, as some are cut. One might perhaps add 5% to the final answer.

(4) The size of the metric brick is more accurately 225 mm × 112·5 mm × 75 mm, but this does not allow for the thickness of mortar.

5. The Box

In assessing the volume of material used in making a box, there is a point to bear in mind when thickness of material cannot be neglected. If the box has a lid, there are two thicknesses of material in each direction. If there is no lid, there is only one thickness in a vertical direction (assuming the box is standing on its base).

We would normally know the external dimensions of the box and the thickness of material.

Consider the lidless box shown. The external length, breadth and height are L, B, H respectively.

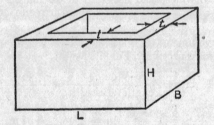

Suppose the thickness of the wood used is t (*measured in the same units as L, B and H*).

External length   = L, Internal length   = $L-2t$
External breadth = B, Internal breadth = $B-2t$
External height   = H, Internal height   = $H-t$

∴ External volume = LBH

Internal volume   = $(L-2b)(B-2t)(H-t)$

∴ Volume of material used

$$= \mathbf{LBH-(L-2t)(B-2t)(H-t)}.$$

If the box had had a lid, then
Volume of material used

$$= \mathbf{LBH} - (\mathbf{L} - 2t)(\mathbf{B} - 2t)(\mathbf{H} - 2t)$$

where H in this case is measured to the top of the lid, when closed.

Ex. 7. Find the volume of material used in making a box, without a lid, of external measurements 23 cm long, 14 cm broad and 9 cm high. The wood is 5 mm thick.

$$L = 23 \text{ cm} \qquad \therefore L - 2t = 22 \text{ cm}$$
$$B = 14 \text{ cm} \qquad B - 2t = 13 \text{ cm}$$
$$H = 9 \text{ cm} \qquad H - t = 8 \cdot 5 \text{ cm}$$

$\therefore$ Volume required

$$= (23 \times 14 \times 9 - 22 \times 13 \times 8 \cdot 5) \text{ cm}^3$$
$$= (2898 - 2431) \text{ cm}^3$$
$$= \mathbf{467 \text{ cm}^3}.$$

EXERCISE 3

1. A garden swimming-pool is 8 metres long and 4·5 metres broad. It contains water to a depth of 1·4 m. Find the mass, in tonnes, of water in the pool (see p. 52), assuming that the bottom of the pool is horizontal.

2. Water flows into a tank 12 m long and 9·45 m broad. How long will it take for the water level to rise by 1 cm, if the rate of flow is 150 litres a minute? (Give the answer correct to the nearest second.)

3. A sheet of metal is 3·2 metres long and 1·25 metres wide. It has a mass of 168 kg. If the metal has a mass of 10·5 grammes per cubic centimetre, find the thickness of the sheet in millimetres.

4. 7 mm of rain fall on a field of area 0·6 hectare. Calculate, in tonnes, the mass of water which has fallen.

5. Find the volume of wood required to make a box of *internal* dimensions 47 cm, 34 cm and 22 cm, if the wood is 1 cm thick, and the box is to have a lid. If the wood costs 60p per square metre surface area, find the cost of wood needed to make the box. (Allow 20% for wastage, i.e. buy 20% more than is used.)

6. A path 36 m long is to be laid in a garden. If the width of the path is 1·2 m, find the cost of covering it with gravel to a depth of 8 cm, taking the cost of gravel to be £3·50 a cubic metre.

6. The Area of a Triangle

Suppose we have triangle ABC (the symbol for triangle is $\triangle$).
Construct a rectangle BCDE on the same base BC and of the same height as $\triangle$ ABC, then DE passes through A. Draw AX perpendicular to BC.

T.Y.A.—7

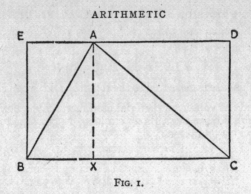

FIG. 1.

The diagonal line AB divides rectangle AEBX into two equal parts.

$$\therefore \triangle ABX = \tfrac{1}{2}\,\text{rect. AEBX}^{15} \qquad . \quad . \quad . \quad . \quad . \quad (1)$$

Similarly $\triangle ACX = \tfrac{1}{2}\,\text{rect. ADCX} \qquad . \quad . \quad . \quad . \quad . \quad (2)$

In fig. 1 *add* these

$$\therefore \triangle ABC = \triangle ABX + \triangle ACX$$
$$= \tfrac{1}{2}\,\text{rect. AEBX} + \tfrac{1}{2}\,\text{rect. ADCX}$$
$$= \tfrac{1}{2}\,\textbf{rect. BCDE.}$$

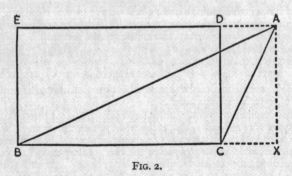

FIG. 2.

In fig. 2 subtract (2) from (1).

$$\therefore \triangle ABC = \triangle ABX - \triangle ACX$$
$$= \tfrac{1}{2}\,\text{rect. AEBX} - \tfrac{1}{2}\,\text{rect. ADCX}$$
$$= \tfrac{1}{2}\,\textbf{rect. BCDE.}$$

Both cases give the same result. Now area of rect. BCDE = BC × DC, but this is the same as BC × AX.

[15] The sign of equality is here taken to mean 'equal in area'.

$\therefore \triangle ABC = \frac{1}{2} BC \times AX$
$= \frac{1}{2} \text{ base} \times \text{height}$
$= \frac{1}{2} \text{ bh}$ (if base $= b$ units, height $= h$ units).

Ex. 8. Find the area of the side of the lean-to greenhouse shown.

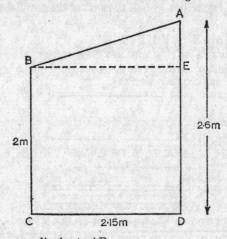

Draw BE perpendicular to AD.
AE $= 2\cdot6$ m $- 2\cdot0$ m $= 0\cdot6$ m

Area of ABCD $= \triangle$ ABE+rect. BCDE
$= \frac{1}{2} \times BE \times AE + BE \times ED$
$= \frac{1}{2} \times 2\cdot15 \times 0\cdot6 + 2\cdot15 \times 2 \text{ m}^2$
$= 0\cdot645 + 4\cdot3$ sq. m. $\cong$ **4·95 m².**

7. The Area of a Trapezium

A *trapezium* is a quadrilateral (four-sided figure) with *one* pair of opposite sides parallel.

In the trapezium ABCD, join AC and draw AX perpendicular to CD produced, and CY perpendicular to AB.

Let AB $= a$, CD $= b$, AX $=$ YC $= h$ (because AB is parallel to DC).

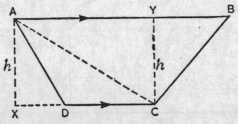

Area of trapezium $= \triangle \text{ABC} + \triangle \text{ADC}$
$$= \tfrac{1}{2} \text{AB} \times \text{CY} + \tfrac{1}{2} \text{DC} \times \text{AX}$$
$$= \tfrac{1}{2} ah + \tfrac{1}{2} bh$$
$$= \tfrac{1}{2}(a+b)\mathbf{h}.$$

∴ **Area of a trapezium = half the sum of the parallel sides × their distance apart.**

8. The Volume of a Right Prism

A right *prism* is a solid of uniform cross-section, the end sections being perpendicular to the generating edges. A greenhouse is prismatic in shape, and so is a girder.

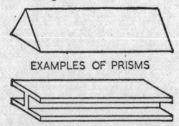

EXAMPLES OF PRISMS

The Volume of a prism = area of cross-section × length.

Ex. 9. Find the area of the given figure, ABCDE. Angles A, C and D are right angles.

Join EB.

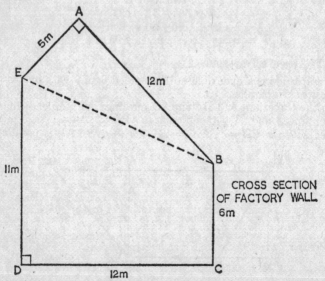

CROSS SECTION
OF FACTORY WALL

Area of figure ABCDE

$= $ Area of $\triangle$ ABE$+$Area of trapezium BCDE
$= \frac{1}{2}$AB$\times$AE$+\frac{1}{2}$(BC$+$ED)$\times$DC
$= \{\frac{1}{2}\times 12\times 5+\frac{1}{2}(6+11)\times 12\}$ m^2
$= \{30+102\}$ m$^2 =$ **132 m^2.**

Notice that we take AB as base of $\triangle$ ABE, so that AE is its height, because AE is perpendicular to AB. Further, ED is parallel to BC because both are perpendicular to DC.

Exercise 4

1. Find the areas of triangles with the following measurements:
   (*a*) base 6 m, height 9 m
   (*b*) base 2·5 cm, height 3·6 cm.

2. A triangle is of area 112 cm^2 and its height is 14 cm. Find the length of its base.

3. Calculate the area of a trapezium in which the parallel sides are 1·72 m and 2·38 m and which are 74 cm apart. Give the answer in square metres.

4. A trough is of the shape shown. ABCD is a horizontal rectangle, and E and F are at a depth 41 cm below AB. Calculate the capacity of the trough in litres. Using the results on p. 50, calculate the number of gallons the trough will hold.

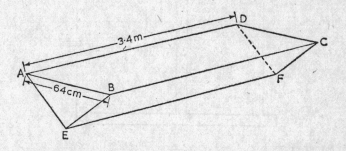

5. The end wall of a house is shown in the diagram overleaf. How many bricks are needed to build the wall, if it is two bricks thick? A brick is 225 mm long, 112·5 mm wide and 75 mm deep and it is laid flat, so that there is 8 cm (including cement or mortar) between successive courses. Take the length of a brick, including mortar, to be 230 mm.[16]

---

[16] In practice, as bricks are not cut with diagonal corners, some extra would be ordered to cover wastage. (See note (3) on page 104.)

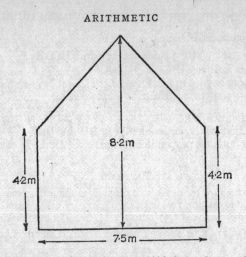

6. Find the mass of a girder made of steel if it is 8 m long and is of cross-sectional area shown. All the angles in the figures are right-angles. Steel has a mass of 7710 kilogrammes per cubic metre (kg/m³).

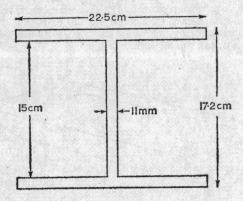

9. Square Root

The *square root* of a given number N is a number y which, when multiplied by itself, gives N. That is

$$y \times y = N$$
$$\text{i.e.} \quad y^2 = N$$

y is the square root of N. It is written

$$\mathbf{y} = \sqrt{\mathbf{N}}.$$

Some numbers have exact square roots, e.g.

$$\sqrt{9} = 3 \text{ (because } 3 \times 3 = 9).$$

The sequence of numbers 1, 4, 9, 16 . . . which are *perfect squares* have exact square roots, 1, 2, 3, 4 . . . .

Most numbers do not have exact square roots. Consider $\sqrt{3}$. $1 \times 1 = 1$, $2 \times 2 = 4$ ∴ $\sqrt{3}$ lies between 1 and 2. Trying again we have $1 \cdot 7 \times 1 \cdot 7 = 2 \cdot 89$ and $1 \cdot 8 \times 1 \cdot 8 = 3 \cdot 24$, so more accurately $\sqrt{3}$ lies between $1 \cdot 7$ and $1 \cdot 8$. Now it would be very tedious to find square roots by trial-and-error methods such as this. (*Teach Yours Books— New Mathematics*, Chapter I, p. 24, develops the idea.) There is, however, a simple way of finding square roots as accurately as we wish.

We will illustrate this method by examples.

Ex. 10. Find $\sqrt{66\ 049}$. (The square root is exact.)

		2	5	7
(i) 2	6	60	49	
	4			
(ii) 45	2	60		
	2	25		
(iii) 507		35	49	
		35	49	
		.....		

∴ $\sqrt{66\ 049} = \mathbf{257}$.

The steps are as follows.

(*a*) Pair off the numbers *both ways* from the decimal point *if there is one*. *If not, pair off from the right-hand end* [e.g. 6 60 49].

(*b*) Find the largest number which, when multiplied by itself, does not exceed the first paired group. (In this case 6.) The number is 2, for $2 \times 2 = 4$. 3 would have been too big, for $3 \times 3 = 9$. Put the 2 on the top line and at (i).

(*c*) Subtract the result 4 from 6. $6 - 4 = 2$. Bring down the next *pair* 60, making 260.

(*d*) Double the 2 at (i) and write the result at (ii). Find the largest number which can be written after the 4 at (ii), and the result multiplied by this new number to give a number smaller than 260. The number is 5, because $45 \times 5 = 225$. The number 6 would have been too big, for $46 \times 6 = 276$, which is bigger than 260. Put the 5 on top and next to the 4 at (ii).

(e) Subtract 225 from 260. 260−225 = 35. Bring down the next *pair* 49, making 3549.

(f) Double the *last figure only* of the 45 at (ii) and rewrite at (iii). Result is 40+5×2 = 50. Find the largest number which can be written after the 50 and the result multiplied by this new number to give a number which will not exceed 3549. The number is 7, because 507×7 = 3549 exactly.

Ex. 11. Find $\sqrt{3}$ correct to 3 decimal places. (The square root is not exact.)

	1·7 3 2 0
I	3·00 00 00 00
	I
27	2·00
	1 89
343	11 00
	10 29
3462	71 00
	69 24
3464	1 76 00

∴ Correct to 3 decimal places, $\sqrt{3} \backsimeq$ **1·732.**

The determination of the position of the decimal point requires care. For the student who is learning from a book of this nature, probably the best method is to make a rough approximation. Suppose we wish to find the square root of 0·447.

Now we get $\sqrt{0·447}$ equal to 668 . . . (by the process already indicated) with a decimal point somewhere; but 0·6×0·6 = 0·36 and 0·7×0·7 = 0·49. ∴ $\sqrt{0·447}$ lies between 0·6 and 0·7. ∴ $\sqrt{0·447}$ = **0·668** . . . .

Multiplication of a number by 100 multiplies the square root by 10. Division of a number by 100 implies division of the square root by 10. Take care to avoid the common mistake $\sqrt{2} \backsimeq 1·4142$. ∴ $\sqrt{0·2} \backsimeq$ 0·141 42. The number 2 has been divided by 10, so that its square root should have been divided by $\sqrt{10}$, i.e. by 3·162. The actual value of $\sqrt{0·2}$ is approximately 0·4472.

Ex. 12. Find $\sqrt{0·00\ 635}$, correct to 3 decimal places.
From the above method:

$$\begin{array}{r|l} & 0\cdot0 \quad 7 \quad 9 \quad 6 \\[4pt] \hline 0 & 0\cdot00\ 63\ 50\ 00 \\ 7 & 49 \\[2pt] \cline{2-2} 149 & 14\ 50 \\ & 13\ 41 \\[2pt] \cline{2-2} 1586 & 1\ 09\ 00 \\ & 95\ 16 \end{array}$$

Check: $0\cdot07 \times 0\cdot07 = 0\cdot0049$; $0\cdot08 \times 0\cdot08 = 0\cdot0064$.

$\therefore \sqrt{0\cdot00\ 635} = 0\cdot0796\ldots \approxeq \mathbf{0\cdot080}$ **(3 d.p.).**

EXERCISE 5

Find the square roots of the following numbers, exactly:

1. 5329   **2.** 925 444   **3.** 9 096 256.

Evaluate the following, correct to 4 significant figures:

4. $\sqrt{2}$ **5.** $\sqrt{16\cdot83}$ **6.** $\sqrt{4\cdot937}$

7. $\sqrt{219\cdot9}$ **8.** $\sqrt{6423}$ **9.** $\sqrt{0\cdot397}$

10. $\sqrt{0\cdot0584}$ **11.** $\sqrt{0\cdot007\ 56}$ **12.** $\sqrt{0\cdot000\ 773}$.

10. Pythagoras's Theorem and Right-angled Triangles

Suppose we have triangle ABC with a right-angle at C. The side opposite the right-angle is called the hypotenuse. Let the length of the sides opposite angles A, B, C be a, b, c units respectively.

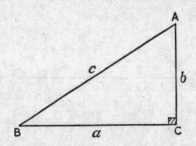

Pythagoras's Theorem states that the square on the hypotenuse of a right-angled triangle is equal to the sum of the squares on the other two sides.

In our diagram, $c^2 = a^2 + b^2$.
If, for example, $a = 4$ cm, $b = 3$ cm
then $c^2 = 4^2 + 3^2$
$= 16 + 9 = 25$
$\therefore c = \sqrt{25} = 5$ **cm.**

A right-angled triangle such as this, wherein the three sides are integral, i.e. are whole numbers, is called Pythagorean.

Other examples of Pythagorean triangles are:

(a) 5, 12, 13: for $5^2 + 12^2 = 169 = 13^2$
(b) 8, 15, 17: for $8^2 + 15^2 = 289 = 17^2$.

Ex. 13. Find the longest side of a right-angled triangle in which the other sides are 1 m and 2 m.

Let the length be x.

$$\therefore x^2 = 1^2 + 2^2 = 5$$
$$20 \quad x = \sqrt{5} = 2.236 \text{ m.}$$

Ex. 14. Find a in the given figure, in which AC = 15 m, BC = 11 m and angle ABC = 90°.

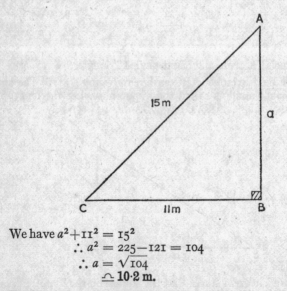

We have $a^2 + 11^2 = 15^2$
$\therefore a^2 = 225 - 121 = 104$
$\therefore a = \sqrt{104}$
$\triangleq 10.2 \text{ m.}$

Ex. 15. A lean-to greenhouse has end sections as shown. The shaded area indicates the height to which bricks are used in its construction, the rest being covered with glass in steel supports. The greenhouse is 3 m long. Calculate the volume of air in the greenhouse. Also find the area of glass required in its construction, neglecting the thickness of the steel.

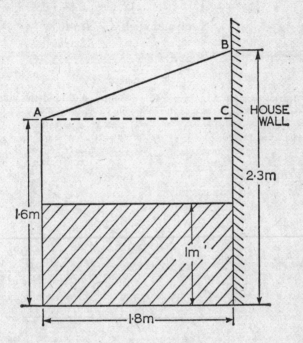

Cross-sectional area of greenhouse $= \frac{1}{2}(1\cdot6 + 2\cdot3) \times 1\cdot8$ m^2
∴ Its volume $\qquad = \frac{1}{2} \times 3\cdot9 \times 1\cdot8 \times 3$ m^3
$\qquad\qquad\qquad\qquad = \mathbf{10\cdot53\ m^3}.$

Area of glass $=$ Area of front$+2\times$Area of end$+$Sloping Roof Area.

Now AB is found by Pythagoras's Theorem, by drawing AC perpendicular to the wall.

$$AB^2 = AC^2 + BC^2 = (1\cdot8)^2 + (2\cdot3 - 1\cdot6)^2$$
$$= 3\cdot24 + 0\cdot49 = 3\cdot73$$
$$\therefore AB = \sqrt{3\cdot73} \simeq 1\cdot932\ m.$$

Sloping roof area
$$= 3 \times 1\cdot932 \qquad\qquad = 5\cdot796\ m^2$$

Area of front
$$= 3 \times (1 \cdot 6 - 1) = 3 \times 0 \cdot 6 = \quad 1 \cdot 800 \, \text{m}^2$$
2 × Area of Glass End
$$= 2 \times \tfrac{1}{2}(0 \cdot 6 + 1 \cdot 3) \times 1 \cdot 8 \quad = \quad 3 \cdot 420 \, \text{m}^2$$
$$\therefore \text{Total area of glass} = 11 \cdot 016 \, \text{m}^2$$
$$\simeq 11 \, \text{m}^2.$$

EXERCISE 6

1. In the triangle ABC, angle C is a right-angle. Find the value of the unknown side in each of the following cases:

(a) $a = 20, b = 21$ (b) $c = 37, a = 35$
(c) $a = 11, c = 61$ (d) $a = 45, b = 28$.

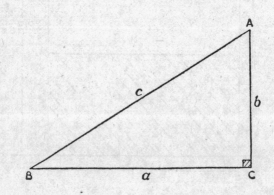

Find the hypotenuses of right-angled triangles, given that the other two sides are:

2. 63, 16   **3.** 3, 7   **4.** 9·4, 16·8
5. 209, 476   **6.** 93·5, 46·8   **7.** 0·86, 0·72.

8. Find the third side of a right-angled triangle in which the hypotenuse is 23 cm and one side is 16 cm.

9. A ladder of length 7 m rests with one end on horizontal ground and the other against a vertical wall. If the distance of the foot of the ladder from the bottom of the wall is 1·4 m find how far up the wall the ladder reaches. (Answer in metres, correct to the nearest centimetre.)

10. The figure ABCD represents the end section of a shed. Angles C and D are right angles. If BC = 2 m, AD = 4·1 m and AB = 3·5 m, calculate CD and the area of ABCD.

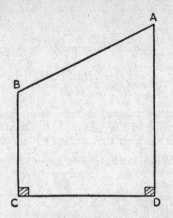

11. Calculate the cost of making the shed in question 10, if it is 4 m long, and if it is to be made entirely of planks 15 cm wide. There is no floor, and the cost of the planks is 12p a metre run.[17]

12. Find the perimeter of a square field of area 0·42 hectare, giving the result to the nearest metre.

13. A pyramid VABCD has a square base ABCD and vertex V. VA = BC = VC = VD = 7 cm, AB = 4 cm. Calculate the height of the vertex above the base, correct to 3 significant figures.

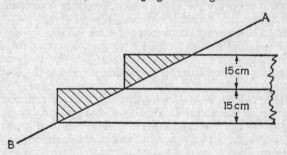

[17] It is worth observing that rather more timber would be needed in practice, to allow for cutting end sections of shed along line AB.

CHAPTER TWELVE

LOGARITHMS

1. Indices

In Chapter 3 brief reference was made to the idea of an index. It is now necessary to investigate this concept a little more closely.

Consider 4×8. We know that $4 = 2^2, 8 = 2^3$
$$\therefore\ 4 \times 8\ = 2^2 \times 2^3 = 2.2.\ 2.2.2$$
$$= 2^5$$
Also $27 \times 81 = 3^3 \times 3^4 = 3.3.3.\ 3.3.3.3$
$$= 3^7.$$

In each of the above, the final index is found to be the sum of the original indices: $2+3 = 5; 3+4 = 7$.

This suggests that when we multiply two numbers together, if we could express them as *powers* of some other number (powers of 2 or 3 in the above examples), then we could replace the process of *multiplying* the original numbers by the simpler process of *adding* the *indices* of the new numbers.

Now this is not a very helpful idea unless there is some easy way of (a) converting the original numbers to powers of some other number (called the *base*) and (b) converting the result (2^5 or 3^7 in the above examples) back into an ordinary number. Fortunately, however, there exist sets of tables for this very purpose. The method by which they are constructed is quite simple but is outside the scope of this book.

The *base* which we use in logarithmic work in arithmetic is the number 10, as this accords with our decimal system of counting. It is not, however, the only base in general use,[18] but it will suffice for our needs. Logarithms to the base 10 are called common, or Briggian, logarithms.

$$\text{consider } 10^2 \times 10^3 = 10^5.$$
$$\text{Now } 10^2 = 100,\ 10^2 = 1000,\ 10^5 = 100\ 000.$$

2, 3, 5 are called the logarithms of 100, 1000, 100 000 respectively to the base 10. They can be written

[18] Logarithms are first calculated, in making up tables, to the base $e = 2 \cdot 718\ 28\ldots$. Logarithms to the base e are called Naperian logarithms, after their inventor, Napier.

$$2 = \log_{10} 100, \; 3 = \log_{10} 1000, \; 5 = \log_{10} 100\,000.$$

We now need 4 important formulæ. If x and y are any numbers,

$$10^x \times 10^y = 10^{x+y} \quad . \quad . \quad . \quad . \quad (1)$$
$$10^x \div 10^y = 10^{x-y} \quad . \quad . \quad . \quad . \quad (2)$$
$$(10^x)^y = 10^{xy} \quad . \quad . \quad . \quad . \quad (3)$$
$$\sqrt[y]{10^x} = 10^{x/y} \quad . \quad . \quad . \quad . \quad (4)$$

We shall not attempt the general proof of these formulæ but shall verify them for a few simple cases.

We have already seen that $10^2 \times 10^3 = 10^5$ which verifies (1).

Now $10^6 \div 10^4 = \dfrac{10.10.10.10.10.10}{10.10.10.10} = 10^2$ which verifies (2) as $6-4 = 2$.

$(10^3)^2 = (10.10.10)(10.10.10) = 10^6$ which verifies (3) as $3 \times 2 = 6$.

$\sqrt[3]{10^6} = \sqrt[3]{(10.10)(10.10)(10.10)} = 10^2$ which verifies (4) as $\frac{6}{3} = 2$.

There is a particular case of great importance. It is the value to be assigned to 10^0.

Now by (1), $10^2 \times \boxed{10^0} = 10^{2+0} = 10^2$ if (1) is to hold in this case,

but $10^2 \times \boxed{1} = 10^2$ as multiplication by one does not alter the value of a quantity.

$\therefore$ 10^0 must be defined as 1.

We then have the following basic table.

Number	Logarithm (to base 10)	
1	0	
10	1	. . . (5)
100	2	
1000	3	
10000	4	
		etc.

We observe that the *logarithm* of each of these numbers is one less than the number of whole *digits* in it. This is important, as will be seen later.

So far we have confined ourselves to numbers which are whole powers of ten. We must now extend our ideas to cover intermediate numbers.

2. Graphical Considerations

Let us construct a graph of the function $y = 10^x$, i.e. the graph in which x is the logarithm of y to the base 10.

Using table (5) above, and plotting the values of x (the *logarithms* 0, 1, 2, 3 . . .) along the horizontal axis, and the values of y (the *numbers* 1, 10, 100 . . .) along the vertical axis, we get a set of points (0, 1), (1, 10), (2, 100), (3, 1000) as shown. It is necessary to use different scales for x and y because of the rapid increase in y.

Join up the points by a smooth curve, and we can then read off the logarithms of intermediate values.

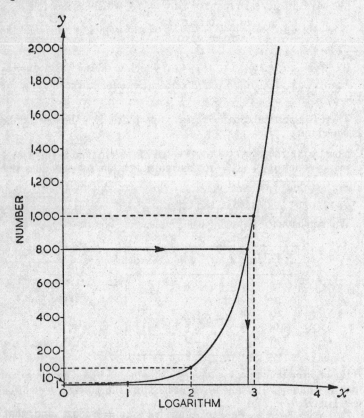

For example, log 800 ≏ 2·9, that is 800 = $10^{2\cdot9}$.
The graph clearly shows that *every number between*:

    1 and   10 has a logarithm between 0 and 1;
    10 and   100 has a logarithm between 1 and 2;
   100 and 1000 has a logarithm between 2 and 3 etc.

Now $800 = 8 \times 100 = 10^2 \times 8$, so if $8 = 10^x$, we can see that $10^2 \times 10^x = 10^{2 \cdot 9}$. Therefore $x = \textbf{0·9}$.

∴ The logarithm of 8 differs from the logarithm of 800 only by its whole number part. This idea is perfectly general:

The logarithm of any number consists of:

—(1) A whole number part, called the *characteristic*, which depends only on the number of figures before the decimal point in the given number;

(2) A decimal part, called the *mantissa*, which depends on the actual figures present, irrespective of the position of the decimal point.

3. Logarithms

Turn to the table of logarithms, part of which is reconstructed below.

2·313	A	B	C		P	Q	R	
No.	Log	1	2	...	1	2	3	...
2·3	·3617	**3636**	3655		2	4	**6**	
2·4	·3802	3820	3838		2	4	5	
2·5	·3979	3997	4014		2	3	5	

Suppose we require log 2·4. We read 2·4 down the left-hand side. In the column under Log, headed A, *and opposite* 2·4 we read 3802.

$$\therefore \log 2 \cdot 4 = 0 \cdot 3802.$$

Similarly if we had wanted log 24 we would have used the same *mantissa* (·3802), but a new *characteristic* (1), determined by the rule in section 2 above, because 24 lies between 10 and 100.

$$\therefore \log 24 = 1 \cdot 3802.$$

Proceeding thus, we see that log 240 = 2·3802, and that log 2400 = 3·3802 and so on.

When the given number has three significant figures in it, such as

2·42, we read the first two numbers, as before, down the left-hand side; but now read off the result under the column, headed C, for the third figure (2) in 2·42.

∴ log 2·42 = 0·3838, log 24·2 = 1·3838, and similarly log 242 = 2·3838 etc.

If the number were, say, 2·51, we would read 2·5 down the side and pick out ·3997 under the column for 1 (headed B). It follows that log 2·51 = 0·3997.

Finally, when the given number has four significant figures, such as 2·313, we read off the first two down the left-hand side and the third in its appropriate column (in this case 1, headed B). This gives 0·3636. The fourth figure is entered in the *mean difference column*, in the same row as before, and under the column for this fourth figure (in this case 3, column R) we read off 6.

So we have

	No.	*Log*
	2·313	0·3636
		+6
		0·3642.

We add the figure obtained (in the above it is 6) to the last figure in the mantissa, as shown. The arrows in fig. 2 should make the procedure clear.

A brief explanation may be helpful.

$$\left.\begin{array}{l} \log 2·31 = 0·3636 \\ \log 2·32 = 0·3655 \end{array}\right\} \text{The difference is 0·0019.}$$

Now 2·313 is $\frac{3}{10}$ of the way from 2·31 to 2·32. We may therefore conjecture that log 2·313 is very near $\frac{3}{10}$ of the way from 0·3636 to 0·3655; but $\frac{10}{10}$ represents a move of 0·0019.

∴ $\frac{3}{10}$ represents a move of $\frac{3}{10} \times 0·0019 = 0·00\ 057 \doteq 0·0006$ to 4 decimal places, and this is the figure which we obtained from the mean difference column.

This process of estimating the difference between one number and another and calculating a fraction of this difference is called *interpolation*. It is a powerful mathematical method, widely used in statistical and actuarial work.

A few examples are now shown, using the logarithm table on pages 213–214. The student should verify them for himself. He should learn to add differences (where necessary) mentally as soon as possible.

No.	Log	No.	Log
4·7	0·6721	8·647	0·9369
8	0·9031	92·36	1·9655
2·97	0·4728	104·9	2·0207

EXERCISE I

1. Find the logarithms of the following numbers:
   - (a) 6
   - (b) 9·2
   - (c) 31
   - (d) 1·64
   - (e) 7·25
   - (f) 6·48
   - (g) 86·2
   - (h) 19·7
   - (i) 4·845
   - (j) 32·41
   - (k) 317·2
   - (l) 1·048
   - (m) 164 000
   - (n) 9·897
   - (o) 5450
   - (p) 70·98.

2. Draw the graph of $y = \log x$ from $x = 1$ to $x = 10$, taking the values of x as follows:

x	1	2	3	4	5	6	7	8	9	10
$y = \log x$	0	·3010	·4771	·6021						1

Firstly, complete the above table, then, taking the x-axis as horizontal, mark off the values of x one centimetre apart on graph paper.

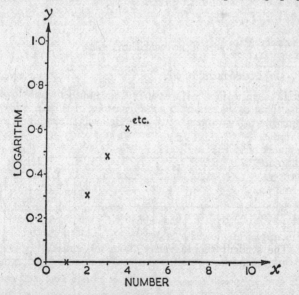

Mark each centimetre on the y-axis as 0·1 units as indicated in the sketch. Join the points obtained from the table above by a smooth curve.

When completed, from your graph read off the values of:[19]

$$\log 6·5, \log 3·7, \log 8·3;$$

and also the numbers whose logs are 0·42, 0·59, 0·20.

Check your results from the table of logarithms.

4. The Four Rules of Logarithms

The four formulæ (1), (2), (3), (4) given in section 1 above constitute the four rules of logarithms.

I. When we wish to *multiply* two numbers, we *add* their logarithms.

II. When we wish to *divide* one number by another, we *subtract* the logarithm of the denominator from that of the numerator.

III. When we wish to find the yth power of a number, we *multiply* the logarithm by this power, y.

IV. When we wish to find the yth root of a number, we *divide* the logarithm by y.

Some examples will soon make this clear, but we must first find how to use the *antilogarithm table*.

Suppose we wish to find $3·624 \times 18·5$.

No.	Log	
3·624	0·5592	} from the logarithm table
18·5	1·2672	
+	**1·8264**	using rule I.

Now this is only of value if we can find the number whose logarithm is 1·8264. That is, we require the *antilogarithm* of 1·8264. Turn to the antilogarithm tables on pages 215–216. Below is the required part for this example.

·8264											
Log	0	...	5	6	...	1	2	3	4	...
·81	6457		6531	6546		2	3	5	6	
·82	6607		6683	**6699**		2	3	5	**6**	
·83	6761		6839	6855		2	3	5	6	

[19] It should be possible to get the result accurate to 2 decimal points.

In using the antilogarithm table, *only the mantissa is entered*, i.e. 0·8264. Remember that the characteristic, 1 in this example, only determines the position of the decimal point in the answer. We enter the table in exactly the same way as before in all other respects; ·82 down the left-hand side, 6 along the top, 4 in the difference column, as indicated by the arrows.

Entry ·8264; antilog is 6699+6 = 6705.

To determine the position of the decimal point, note that by section 2 above 1·8264 is the logarithm of a number between 10 and 100.
∴ Antilog 1·8264 = **67·05**, and this is the required result.

EXERCISE 2

1. Find the antilogarithms of the following:

(a) 0·6154	(b) 0·3047	(c) 0·8995	(d) 0·0019
(e) 1·2700	(f) 1·0809	(g) 2·4677	(h) 4·6021
(i) 0·9879	(j) 0·7788	(k) 3·4186	(l) 5·5109.

Ex. 1. Find the value of $1·986 \times 2·324$.

No.	Log	Rule I.
1·986	0·2980	
2·324	0·3662	
+	0·6642 Antilog 0·6642 = **4·615.**	

Ex. 2. Find the value of $327·9 \div 71·8$.

No.	Log	Rule II.
327·9	2·5157	
71·8	1·8561	
−	0·6596 Antilog 0·6596 = **4·566.**	

Ex. 3. Calculate $\dfrac{3·947 \times 86·43}{93·77}$

No.	Log	
3·947	0·5963	Rule I
86·43	1·9367	
+	2·5330	Rule II
93·77	1·9720	
−	0·5610 Antilog 0·5610 = **3·639.**	

Ex. 4. Evaluate $(3·942)^4$.

No.	Log	
3·942	0·5957	Rule III
×4	2·3828 Antilog 2·3828 = **241·4.**	

Ex. 5. Find the cube root of 71·2.

No.	Log
71·2	1·8525
÷3	0·6175

Rule IV

÷3 | 0·6175 ∴ $\sqrt[3]{71·2} = $ **4·145.**

For emphasis, we repeat the following rules:

A. Numbers between   1 and   10 have *characteristic* 0
   „   „   10 and 100   „   „   1
   „   „   100 and 1000 „   „   2 etc.

B. Only the *mantissa* is used in the table itself.

$$\log 63·8 = \textbf{1·8048}$$

CHARACTERISTIC_____↑ ↑_____MANTISSA

EXERCISE 3

Find the values of the following expressions:

1. 3·47 × 9·18
2. 61·3 × 2·054
3. 322 × 2·76
4. 1·048 × 32·95
5. 3·646 × 1·922
6. 307·4 × 296·6
7. 38·4 ÷ 9·77
8. 6·029 ÷ 5·73
9. 3724 ÷ 717
10. 3985 ÷ 21·76
11. 36·88 ÷ 9·828
12. 67 500 ÷ 43·84
13. $\dfrac{6724 \times 1·925}{837·1}$
14. $\dfrac{3·28 \times 6·95}{12·74}$
15. $\dfrac{214·6 \times 38·7}{61·5 \times 9·212}$
16. $\sqrt{24·9}$
17. $\sqrt[5]{8·637}$
18. $(21·95)^3$
19. $(6·009)^4$
20. $\sqrt[4]{(17·28)^3}$ (multiply the log by 3 and divide the result by 4).

5. Negative Characteristics

The student will probably have noticed that hitherto we have restricted ourselves to numbers greater than 1. Suppose, however, we require the logarithm of a number less than 1, say 0·6248. Before we can discuss this in detail it is necessary to consider powers of ten again.

Rule (1) in Section 1 gave us

$$10^x \times 10^y = 10^{x+y}$$

Suppose now that $x = 4$ and $y = -1$.

$$\therefore 10^4 \times \boxed{10^{-1}} = 10^3$$
$$\text{but we know } 10^4 \times \boxed{\tfrac{1}{10}} = 10^3$$

so therefore 10^{-1} and $\tfrac{1}{10}$ must be the same quantity.

We infer that $10^{-1} = \frac{1}{10} = 0·1$

$$10^{-2} = \frac{1}{100} = 0·01$$
$$10^{-3} = \frac{1}{1000} = 0·001 \text{ etc.}$$

∴ **The logarithms of numbers which lie between 0 and 1 are always negative.**

Now we know that $0·6248 = 6·248 \times \frac{1}{10}$.

∴ log $0·6248 =$ log $6·248 +$ log $\frac{1}{10}$, but we have seen that the log $\frac{1}{10} = -1$ from above.

∴ log $0·6248 = 0·7958 - 1$
$$= \bar{1}·7958.$$

We put the minus over the characteristic to show that this is negative, but that the mantissa is not. It is really a short way of writing $-1 + ·7958$.

Similarly log $0·06248 = 0·7958 - 2$
$$= \bar{2}·7958$$
$$\log 0·006248 = 0·7958 - 3$$
$$= \bar{3}·7958 \text{ and so on.}$$

We can now extend rule A in section 4 above:
Numbers between

100 and 1000 have characteristic 2 etc. ....
  10 and 100 ,,   ,,   1
   1 and 10 ,,   ,,   0
  0·1 and   1 ,,   ,, $\bar{1}$
 0·01 and 0·1 ,,   ,, $\bar{2}$
0·001 and 0·01 ,,   ,, $\bar{3}$ etc. ....

Ex. 6. Find the logarithms of $0·82$, $0·004\ 71$, $0·005\ 228$.

No.	Log
0·82	$\bar{1}·9138$
0·004 71	$\bar{3}·6730$
0·005 228	$\bar{3}·7184$

EXERCISE 4

1. Find the *characteristics* of the logarithms of the following numbers:
  (a) 3·94   (b) 0·7287   (c) 31 950   (d) 0·000 622
  (e) 0·093 854   (f) 0·000 06   (g) 0·010 08.

2. Find the logarithms of the following numbers:

    (a) 0·2194   (b) 0·063   (c) 21 776   (d) 0·000 952 6

    (e) 0·090 91   (f) 0·8637   (g) 0·000 000 62.

3. Find the antilogarithms of the following:

    (a) 2·645   (b) $\bar{2}$·645   (c) $\bar{1}$·9837   (d) 4·727

    (e) 6·9   (f) $\bar{3}$·9289   (g) $\bar{1}$·0099   (h) $\bar{7}$·7687.

Manipulation with negative characteristics requires care. Their difficulties are best illustrated by examples.

Ex. 7. Multiply 0·6284 by 0·739.

No.	Log
0·6284	$\bar{1}$·7983
0·739	$\bar{1}$·8686
+	$\bar{1}$·6669
	1

∴ 0·6284 × 0·739 = **0·4644.**

The point to observe here is that the carrying figure to the characteristics column is positive, whilst the characteristics are themselves negative.

∴ $\bar{1}+\bar{1}+1 = \bar{1}$, for this is the same as

$-1-1+1 = -\bar{1}.$

Ex. 8. Multiply 0·0094 by 21·3.

No.	Log
0·0094	$\bar{3}$·9731
21·3	1·3284
+	$\bar{1}$·3015
	1

Antilog $\bar{1}$·3015 = **0·2002.**

Here, $\bar{3}+1+1 = \bar{1}.$

Ex. 9. Evaluate 6·205 ÷ 0·8381.

No.	Log
6·205	0·^{1}7927
0·8381	$_{1}\bar{1}$·9233
−	0·8694

Antilog 0·8694 = **7·403.**

We borrow +1 from the bottom characteristic so that we can get 1·7−0·9 = 0·8. Paying it back makes the bottom characteristic −1+1 = **0.**

∴ Final characteristic = 0−0 = 0.

(Alternatively, putting it in full,

$$0 \cdot 7927 - (-1 + 0 \cdot 9233) = 0 \cdot 7927 + 1 - 0 \cdot 9233$$

on changing the sign of everything inside the bracket, giving $1 \cdot 7927 - 0 \cdot 9233 = 0 \cdot 8694$.)

Ex. 10. Find the value of $(0 \cdot 862)^3$.

No.	Log
0·862	$\bar{1} \cdot 9355$
×3	$\bar{1} \cdot 8065$ ∴ $(0 \cdot 862)^3 = \mathbf{0 \cdot 6404}$.
2	

In this case $3 \times \bar{1} = \bar{3}$, add 2, giving $\bar{1}$.

Ex. 11. Find the value of $\sqrt[4]{0 \cdot 0729}$

No.	Log
0·0729	$\bar{2} \cdot 8627$
÷4	$\bar{1} \cdot 7157$ ∴ $\sqrt[4]{0 \cdot 0729} = \mathbf{0 \cdot 5196}$.

We cannot divide 4 into $\bar{2}$ and get a whole number, so we proceed as follows:

$$\bar{2} \cdot 8627 \div 4 = (\bar{4} + 2 \cdot 8627) \div 4 = \bar{1} + 0 \cdot 7157.$$

The characteristic is always made into an exact multiple of the divisor.

Ex. 12. Evaluate $\sqrt{\dfrac{31 \cdot 46 \times 7 \cdot 92}{3940 \times 0 \cdot 86}}$.

No.	Log
31·46	1·4977
7·92	0·8987
+	2·3964
→	3·5300
−	$\bar{2} \cdot 8664$
÷2	$\bar{1} \cdot 4332$ →Antilog $\bar{1} \cdot 4332 = \mathbf{0 \cdot 2711}$.
3940	3·5955
0·86	$\bar{1} \cdot 9345$
+	3·5300

We (1) add the numerator logs; (2) add the denominator logs; (3) subtract the second result from the first; (4) divide the result by two to get the square root of the original expression.

EXERCISE 5

Find the values of the following expressions:

1. $3 \cdot 82 \times 0 \cdot 463$

2. $0 \cdot 89 \times 0 \cdot 414$

3. $2 \cdot 68 \times 0 \cdot 19 \times 3 \cdot 045$

4. $0 \cdot 0024 \times 0 \cdot 4891$

5. $(0 \cdot 938)^2$

6. $\sqrt{0 \cdot 097\ 45}$

7. $(2 \cdot 91)^2 \times 0 \cdot 003\ 635$

8. $(0 \cdot 0912)^2 \div 0 \cdot 8477$

9. $6 \cdot 328 \div 17 \cdot 914$

10. $0 \cdot 915 \div 214 \cdot 3$

11. $0 \cdot 003\ 85 \div 0 \cdot 001\ 767$

12. $164 \div (0 \cdot 928)^2$

13. $\dfrac{61 \cdot 42 \times 3 \cdot 098}{2 \cdot 94 \times 0 \cdot 8087}$

14. $\dfrac{21 \cdot 38}{(0 \cdot 645)^2 \times 319 \cdot 7}$

15. $\sqrt{\dfrac{0 \cdot 625}{0 \cdot 38 \times 5 \cdot 752}}$

16. $\sqrt[3]{(0 \cdot 8228)^4}$.

6. Harder Problems and Applications to Formulæ

The student will by now have realised the power of logarithmic work to simplify difficult calculations. There are, however, certain pitfalls to be avoided.

$$\text{Consider } \frac{3 \cdot 624}{8 \cdot 041} + \frac{0 \cdot 283}{0 \cdot 914}.$$

The difficulty about this seemingly harmless expression is the *plus* between the two quotients. We have to work out each quotient separately, and to make it perfectly clear the example is worked below.

No.	Log		No.	Log
3·624	0·5592		0·283	$\bar{1}$·4518
8·041	0·9054		0·914	$\bar{1}$·9609
—	$\bar{1}$·6538			$\bar{1}$·4909

Antilog $\bar{1}$·6538   0·4506
Antilog $\bar{1}$·4909   0·3096

+   0·7602.

The antilogarithm of each part must be taken before the final addition. The same care must be exercised when there is a *minus* between two expressions.

Let us now consider the accuracy of our answers. The tables which we have been using are four-figure tables. We have to appreciate therefore that the fourth figure is not usually exactly correct, but is an approximation. For example, $\log 2 = 0{\cdot}30103\ldots$ is given in the tables as $\log 2 = 0{\cdot}3010$. If, however, we work out 2^3 by logs we get the following:

No.	Log
2	0·3010
×3	0·9030⟶Antilog 0·9030 = 7·998.

Now we know perfectly well that $2^3 = 8$, so there is an error of 0·002 in the answer given by the logarithm tables. If we corrected the answer to *three significant figures* we should get $7{\cdot}998 \fallingdotseq 8{\cdot}00$, which is correct.

We can, therefore, expect that, in using four-figure tables, our answer will be *correct to three significant figures*, except when long and involved expressions or high powers (e.g. 3^{14}) occur. In this case the answer may be accurate only to two significant figures.

From now on we shall carry out all working to four significant figures and correct the answer to three significant figures.

Formulæ form an integral part of the modern world, especially in engineering and scientific work. Many people need to gain speed and accuracy in calculating certain results by substituting known values of given quantities in a formula and evaluating the expression so obtained. Four possible methods can be considered:

(1) Long calculation. This has been discussed in detail earlier.
(2) Logarithms.
(3) Slide Rule. Small books of explanation are usually provided at the time of purchase, but the slide rule is now virtually obsolete.
(4) Computer and Electronic Calculator. The basic arithmetic is referred to in the last chapter of this book, but whereas the use of an electronic calculator is fairly simple, that of a computer necessitates special training.

Logarithms are more accurate than slide-rules. Four-figure tables are not the only ones in general use. Inman's Nautical Tables, used in the Royal Navy for navigational work, are five-figure tables. Norie's

Tables, used in the Merchant Service, are six-figure tables. There are sets of seven-figure tables in common use, e.g. Chambers' Mathematical Tables. There are also tables of greater accuracy.

Ex. 13. The time of oscillation of a compound pendulum is given by $T = 2\pi\sqrt{\dfrac{k^2}{lg}}$. Find T when $l = 3\cdot5, g = 9\cdot81, k = 2\cdot94$.

We have

$$T = 2\times\pi\times\sqrt{\frac{(2\cdot94)^2}{3\cdot5\times9\cdot81}}$$

$$= \frac{2.\pi.(2\cdot94)}{\sqrt{3\cdot5\times9\cdot81}}$$

$$= 3\cdot15.$$

(*Note.*—The final log, 0·4985, was obtained by subtracting (ii) from (i).)

No.	Log
2	0·3010
π	0·4971
2·94	0·4683
+	1·2664 (i)
3·5	0·5441
9·81	0·9917
+	1·5358
÷2	0·7679 (ii)
−	0·4985

EXERCISE 6

1. Evaluate $\dfrac{2\cdot71}{6\cdot38}-\dfrac{4\cdot197}{7\cdot925}$.

2. Simplify $\sqrt[3]{\dfrac{0\cdot925\times32}{761\times0\cdot008\,472}}$.

3. If $T = 2\pi\sqrt{\dfrac{l}{g}}$, find T when $l = 1\cdot6, g = 9\cdot81$.

4. The formula $v^2 = u^2+2fs$ occurs in the study of dynamics.
   Find (i) v, if $u = 17\cdot3, f = 0\cdot62, s = 86$
   (ii) s, if $u = 0, v = 38\cdot4, f = 0\cdot93$
   (iii) f, if $u = 61\cdot4, v = 80, s = 7\cdot75$.

5. The area of a triangle is given by the formula
   $\Delta = \sqrt{s(s-a)(s-b)(s-c)}$, where a, b, c are the lengths of the sides and s is the semi-perimeter (i.e. $s = \frac{1}{2}(a+b+c)$). Find Δ, if $a = 7$, $b = 9\cdot4, c = 8\cdot6$.

6. The formula for a gas expanding under adiabatic conditions is $p = \dfrac{k}{v^{1\cdot4}}$, where p is the pressure and v is the volume.
   Find (i) k if $p = 10$ when $v = 6\cdot5$
   (ii) p when $v = 8\cdot74$ using the value of k just found.

7. The focal length of a lens is given by the formula

$$\frac{1}{f} = \frac{1}{v} + \frac{1}{u}$$

Find f if $u = 16 \cdot 3$, $v = 21 \cdot 8$.

8. Simplify $\dfrac{(61 \cdot 35)^2 - (42 \cdot 08)^2}{794 \cdot 6}$.

CHAPTER THIRTEEN

THE CIRCLE, CYLINDER, CONE AND SPHERE

1. Definitions: Properties of a Circle

A *circle* is a path traced out on a plane by a point moving so that its distance from a fixed point, called the *centre*, is always constant. This distance is called the *radius*, *r*, of the circle.

A *chord* of a circle is a straight line, such as CD in fig. 1, which terminates at the circle at both ends.

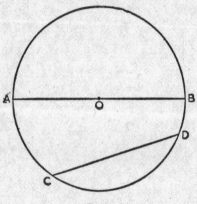

Fig. 1

A *diameter* is a chord passing through the centre O. AB is a diameter, of length *d*, say. All diameters of a circle are equal and they are all bisected at O.

Now OA = OB = *r*, as both are radii.

$$\therefore AB = OA + OB = 2r.$$

The *circumference* of a circle is the distance round its edge. It is equivalent to the perimeter of a polygon. We shall use C to stand for circumference.

One of the earliest references to an important property of the circle occurs in the Bible (2 Chronicles iv, 2) and reads as follows:

'Also he made a molten sea of ten cubits from brim to brim, round in compass, and five cubits the height thereof; and a line of thirty cubits did compass it round about.' The cubit was a length of about 50 cm.

Now this gives a circumference of 30 cubits for a circle of diameter 10 cubits, and leads to an early estimate of the ratio $\dfrac{\text{Circumference}}{\text{Diameter}}$ as being 3. *The ratio, circumference divided by diameter, is constant for all circles,* but the value 3 is a very rough approximation. More accurately, it is $3\frac{1}{7}$ (or $\frac{22}{7}$), a value often used today when great accuracy is not needed. The exact value of the ratio is π (pi). It is an incommensurable quantity, that is, one which cannot be estimated exactly. Very accurate approximations, however, have been made to hundreds of places of decimals.

Here are some approximations:

$$\pi = \tfrac{22}{7},\ \pi = 3\cdot142,\ \pi = 3\cdot1416.$$

Actually $\pi = 3\cdot141\ 592\ 653\ 589\ 793\ 238\ 46\ \ldots$ to 20 decimal places.

Archimedes devised an ingenious method of approximation for π. He inscribed and circumscribed 96-sided regular polygons round a circle, and pointed out that the circumference of the circle lay between the perimeters of the polygons in length. This led him to deduce that π lay between $3\frac{1}{7}$ and $3\frac{10}{71}$.

There is an infinite series of terms which would give π exactly if we could sum the series. It is

$$\frac{\pi}{4} = 1 - \frac{1}{3} + \frac{1}{5} - \frac{1}{7} + \cdots$$

and is called Gregory's series, but it should be pointed out that many terms would have to be taken to give a reasonable value of π. This formula can be found easily by Integral Calculus. Other series giving π also exist.

In this book we shall take π as either $\frac{22}{7}$ or $3\cdot142$ according to our requirements. The former is often used when logarithms are not needed in the question, and the latter is used otherwise.

Now, $\dfrac{\text{Circumference}}{\text{Diameter}} = \pi.$

∴ **Circumference, C** $= \pi d = 2\pi\mathbf{r}$ (because $d = 2r$).

The area of a circle is $\pi\mathbf{r}^2$, that is, π multiplied by the square of the radius. This formula can also be obtained by calculus.

Akmes, an Egyptian, found quite a good approximation to the

area of a circle, nearly 4000 years ago. He suggested that $\frac{8}{9}$ of the diameter be taken and the result squared. This is equivalent[20] to taking π as 3·1605 instead of 3·1416.

Now, area of circle, $A = \pi r^2 = \pi \left(\dfrac{d}{2}\right)^2$, (as $r = \frac{1}{2}d$)

$$= \frac{\pi d^2}{4}.$$

So we can obtain the circumference and area of a circle, given either radius or diameter, but the student is advised always to use $C = 2\pi r$, $A = \pi r^2$ to avoid mistakes. He should therefore convert diameter to radius wherever necessary.

Ex. 1. Find the circumference and area of a circle of radius 3·5 cm. (Take $\pi = \frac{22}{7}$.)

$$r = 3\tfrac{1}{2} \text{ cm} \quad \therefore C = 2\pi r$$
$$= 2 \times \tfrac{22}{7} \times \tfrac{7}{2} \text{ cm}$$
$$= \mathbf{22\ cm.}$$

$$A = \pi r^2$$
$$= \tfrac{22}{7} \times \tfrac{7}{2} \times \tfrac{7}{2} \text{ cm}^2$$
$$= \mathbf{38 \cdot 5\ cm^2.}$$

When using logarithms, it is very helpful to remember that $\log \pi = 0\cdot4971$. Actually if we look up $\log 3\cdot142$ in the tables we get $0\cdot4972$, which is a little less accurate.

Ex. 2. Find the radius of a circle of area 0·35 m².
We have $\pi r^2 = 0\cdot35$ m²
$$= 3500 \text{ cm}^2$$

$$\therefore r^2 = \frac{3500}{\pi}$$

so $r = \sqrt{\dfrac{3500}{\pi}}$

$$= 33\cdot38$$
$$\eqsim \mathbf{33 \cdot 4\ cm.}$$

No.	Log
3500	3·5441
π	0·4971
—	3·0470
÷2	1·5235

---

[20] $\left(\dfrac{8}{9}d\right)^2 = \dfrac{64}{81}d^2$; comparing with $\dfrac{\pi}{4}d^2$ we have $\dfrac{\pi}{4} = \dfrac{64}{81}$.

$$\therefore \pi = \frac{256}{81} = 3\cdot1605.$$

Ex. 3. A reel of thread is to be 100 m long. If the radius of the reel is 1·5 cm, find how many turns there must be on the reel. (Neglect the thickness of the thread.)

In a problem of this kind we take the radius as the *average* radius, neglecting the thickness of the thread.

	No.	Log	
Length of one turn			
$= 2\pi r = 2\pi \times 1\cdot 5$ cm			
$= 3\pi$ cm	10000	4·0000	(i)
Length of thread	3	0·4771	
$= 100$ m $= 10\,000$ cm	π	0·4971	
$\therefore$ Number of turns	$+$	0·9742	(ii)
$= \dfrac{\text{Length of thread}}{\text{Length of one turn}}$	(i)–(ii)	3·0258	

$$= \frac{10000}{3\pi}$$

$\frown$ **1060 turns.**

EXERCISE I

1. Find the circumferences of the circles whose radii are:
   (a) 14 cm   (b) 10·5 cm   (c) 4 cm   (d) 75 cm   ($\pi = \frac{22}{7}$).
2. Find the areas of the circles whose diameters are:
   (a) 4·2 cm   (b) 1·54 m   (c) 27·5 cm   (d) 5 cm   ($\pi = \frac{22}{7}$).
3. A circle is of area 49 cm². Find its diameter and its circumference. ($\pi = 3\cdot142$.)
4. Find the radius of a circle of area 1840 cm².
5. What is the circumference of a circle of area 2 m²?
6. A piece of rope is 88 m long. How many complete turns can be made round a drum of diameter 75 cm? Neglect the thickness of the rope.
7. A bicycle wheel is of diameter 66 cm. How many turns does it make in travelling one kilometre?
8. In a bicycle there are 48 teeth on the sprocket wheel to which the pedals are attached. There are 18 teeth on the sprocket wheel which is fixed to the back wheel of the bicycle. How many times does the rider have to turn the pedals in travelling 400 m, if a bicycle wheel is of diameter 72·5 cm?

2. The Cylinder

The name cylinder covers a fairly large field of solid bodies. We shall restrict our definition to the *right circular cylinder*.

A cylinder is a solid with circular ends and all its generating lines parallel. All its right sections are circles. It can be generated by re-volving a rectangle round one of its edges (e.g. by revolving rectangle ABCD in fig. 2 about the line AD).

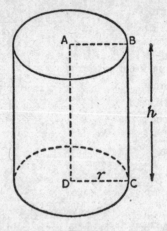

FIG. 2

Let the base radius be r and the height be h, then *the volume* V is given by

$$V = \text{Base area} \times \text{height}$$
$$= \pi r^2 \times h$$
$$= \pi r^2 h.$$

The area of the curved surface of a cylinder is $2\pi rh$. This is easily proved. Suppose we wrap a piece of paper exactly round the cylinder and then unroll it (fig. 3).

The length PR of paper unwrapped
    = Circumference of circle = $2\pi r$.
The height of the paper RS = h.
∴ Area of curved surface of cylinder = Area of PQSR
    = PR × RS = $2\pi r \times h = 2\pi rh.$
The *total surface area of a cylinder* is the sum of the areas of the curved surface and the two ends, so it is given by

$$2\pi rh + 2\pi r^2 = 2\pi r(r+h).$$

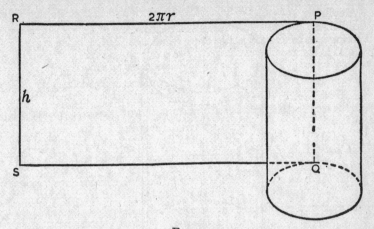

FIG. 3

Ex. 4. Find the volume of a cylinder of diameter 6 cm and height 7 cm. Find also its curved surface area.

We have $r = 3$ cm, $h = 7$ cm

$$\therefore V = \pi r^2 h$$
$$= \tfrac{22}{7} \times 3 \times 3 \times 7 \text{ cm}^3$$
$$= \mathbf{198\ cm^3}.$$

Curved surface area $= 2\pi r h$
$$= 2 \times \tfrac{22}{7} \times 3 \times 7 \text{ cm}^2$$
$$= \mathbf{132\ cm^2}.$$

Ex. 5. Find how many square centimetres of copper are needed to make a can of radius 6·5 cm and height 15 cm. The lid is to overlap the top of the can by 1·6 cm. (Neglect thickness of metal.)

The area of metal required is the same as for a can of height 16·6 cm.

$$\therefore \text{Total surface area} = 2\pi r (r+h)$$
$$= 2\pi \times 6 \cdot 5 \times (6 \cdot 5 + 16 \cdot 6) \text{ cm}^2$$
$$= 2 \times \tfrac{22}{7} \times 6 \cdot 5 \times 23 \cdot 1 \text{ cm}^2$$

$$= 2 \times \frac{22}{\underset{1}{7}} \times \frac{\overset{1}{13}}{\underset{1}{2}} \times \overset{3 \cdot 3}{23 \cdot 1} \text{ cm}^2$$

$$= 943 \cdot 8 \text{ cm}^2$$
$$\simeq \mathbf{944\ cm^2}.$$

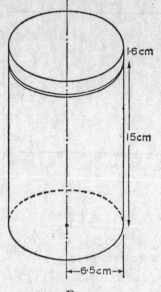

Fɪɢ. 4

3. Material Used in Making a Pipe

A pipe can be thought of as two co-axial cylinders (i.e. cylinders with the same axis). It is easy to find the volume of material used by subtracting the inside volume from the whole volume.

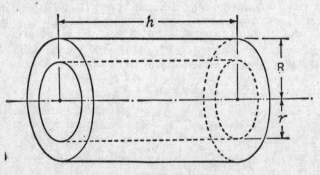

Fɪɢ. 5

Let R, r be the external and internal radii respectively.
Then, volume of material

= Volume of larger cylinder—Volume of smaller cylinder
$= \pi R^2 h - \pi r^2 h$
$= \pi h (R^2 - r^2)$

The student who has done a little algebra will notice that $R^2 - r^2$ can be factorised into $(R+r)(R-r)$, and we have

$$\text{Volume of material} = \pi h (R+r)(R-r).$$

This is the form of answer most useful for logarithmic tables.

Sometimes the external radius, R, and *thickness*, t, are given, in which case $R-r = t$, whence $r = R - t$, giving internal radius.

Ex. 6. Calculate the volume of material used in making a pipe of length 10 m, external radius 1·5 cm and thickness 2 mm. (Take $\pi = \frac{22}{7}$.)

$R = 1\cdot5$ cm, $t = 0\cdot2$ cm. $\therefore r = 1\cdot5 - 0\cdot2 = 1\cdot3$ cm, furthermore $h = 10$ m $= 1000$ cm.

$$\begin{aligned}
\text{Volume of material} &= \pi h (R^2 - r^2) \\
&= \tfrac{22}{7} \times 1000 \times (2\cdot25 - 1\cdot69) \text{ cm}^3 \\
&= \tfrac{22}{7} \times 1000 \times 0\cdot56 \text{ cm}^3 \\
&= \mathbf{1760 \text{ cm}^3} \text{ (approx.).}
\end{aligned}$$

Ex. 7. A crucible is to be made in the form of a cylinder without a lid, the thickness of material being 1 cm throughout. The height is to be 14 cm externally, and the external radius is to be 8 cm. If the material of which it is to be constructed is of mass 7500 kg/m³, find, in grammes, the mass of the crucible. (Log $\pi = 0\cdot4971$.) (See fig. 6.)

The point to bear in mind in this question is that the internal cylinder is of height 1 cm less than the external cylinder, as there is one end present.

If R, H; r, h are external radius and height, internal radius and height respectively, we have $R = 8$ cm, $r = 7$ cm, $H = 14$ cm, $h = 13$ cm.

Volume of material (actually iron)

$= \pi R^2 H - \pi r^2 h$	(a modified formula because
$= \pi [8^2 \times 14 - 7^2 \times 13] \text{ cm}^3$	of the difference in the
$= \pi \times 259 \text{ cm}^3$	heights of the cylinders)

But 1 m³ has mass 7500 kg

$\therefore$ 1 cm³ has mass $7500 \times 1000 \div 1\,000\,000$ gm

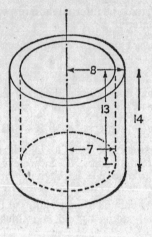

Fig. 6

∴ mass of crucible

$$= \frac{259\pi \times 7\ 500\ 000}{1\ 000\ 000}\ \text{gm}$$

$$= 259\pi \times 7 \cdot 5\ \text{gm}$$

$$\simeq \textbf{6100 gm.}$$

No.	Log
π	0·4971
259	2·4133
7·5	0·8751
+	3·7855

EXERCISE 2

1. A well of diameter 1·5 m has to be excavated to a depth of 5·6 m. Find the volume of earth to be removed. (Answer in cubic metres.)

2. A cylindrical tank of diameter 3 m is full of water. By how much does the water level fall if 2·5 m³ are drawn off?

3. A pipe is of external diameter 16 cm and is made of material 2 cm thick. What is the volume of metal in a 2-metre length of the pipe?

4. A telegraph pole is 6 m high and 25 cm in diameter. What is its mass, if it is of wood of mass 525 kg/m³?

5. Find the total external surface area of a cylindrical can of height 18 cm and diameter 10 cm, if the can has no lid.

6. A zinc trough is made in the form of half a cylinder of diameter 64 cm and length 2·3 m. Neglecting the thickness of the zinc, calculate, in litres, the quantity of water the trough will hold.

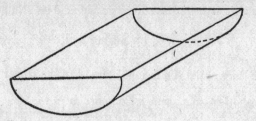

7. Calculate the volume of cupro-nickel used in making a 10p piece. (Make a heap of these coins and stand another one against them, so estimating the average thickness of a 10p piece.)

8. Water flows through a pipe of cross-sectional area 6 cm² at 2·5 m/s. How long will it take to fill a cylindrical tank of base diameter 84 cm and height 85 cm? (Think of cylinders of water 2·5 m long being put into the tank every second.)

4. The Cone

As with the cylinder, we shall limit our definition to the case of a *right circular* cone.

A cone is the solid formed by revolving a line about a fixed axis and always making a constant angle with it. (It can be thought of also as the solid formed by rotating △ VOA about VO in fig. 7.)

A cone has a circular base, centre O in fig. 7, and a *vertex*, V.

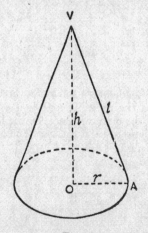

Fig. 7

The cone is a particular case of a pyramid, and its volume is given by:

$$\tfrac{1}{3} \times \text{Area of base} \times \text{height}$$
$$= \tfrac{1}{3} \times \pi r^2 \times h = \tfrac{1}{3}\pi \mathbf{r}^2 \mathbf{h}$$

i.e. it is one-third of the volume of a cylinder of the same radius and height.

There is another length which is important in the cone. It is the *slant height* VA = *l*. Now triangle OAV has a right angle at O, so by Pythagoras's Theorem, we have

$$VA^2 = VO^2 + OA^2$$
$$\text{i.e. } \mathbf{l}^2 = \mathbf{h}^2 + \mathbf{r}^2.$$

The curved surface area of a cone is $\pi \mathbf{r} \mathbf{l}$.

This can be demonstrated by folding a piece of paper exactly once round the curved surface and unrolling it on a table. This gives a sector of a circle.

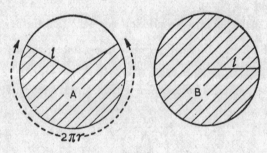

Fig. 8

Suppose we get the shaded area shown on the left, this being part of the circle shown on the right. The circle is of radius *l*. Now the length of arc (part of the circumference) in the left-hand circle was the circumference of base of cone, i.e. $2\pi r$. The length of circumference of the right-hand circle is $2\pi l$.

∴ In fig. 8:

$$\frac{\text{Area of A}}{\text{Area of B}} = \frac{\text{Length of arc of A}}{\text{Length of arc of B}}$$

$$\therefore \frac{\text{Area of A}}{\pi l^2} = \frac{2\pi r}{2\pi l} = \frac{r}{l}$$

$\therefore$ Area of A $= \dfrac{r}{l} \times \pi l^2 = \pi \mathbf{r} \mathbf{l}$, as given above.

(The area of a piece of cake is proportioned to its length of arc.)

Ex. 8. Find the curved surface and volume of a cone of height 4 cm and base radius 3 cm. (Take $\pi = \frac{22}{7}$.)

$$h = 4 \text{ cm}, r = 3 \text{ cm}$$
$$\therefore l^2 = 4^2 + 3^2 = 25 \text{ sq. cm}$$
$$\therefore l = \sqrt{25} = \mathbf{5 \text{ cm}}.$$

$\therefore$ Curved surface $= \pi r l \backsimeq \frac{22}{7} \times 3 \times 5 \text{ cm}^2$
$= \mathbf{47\tfrac{1}{7} \text{ cm}^2}.$

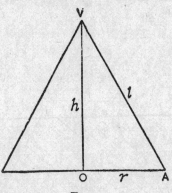

FIG. 9

Volume $\quad = \frac{1}{3}\pi r^2 h$
$= \frac{1}{3} \times \frac{22}{7} \times 3 \times 3 \times 4 \text{ cm}^3$
$= \mathbf{37\tfrac{5}{7} \text{ cm}^3}.$

More Difficult Example

Ex. 9. Find the area of crinothene used in making a lampshade in the form of a frustum of a cone, the upper and lower radii being 7 cm and 12 cm respectively and the slant height being 25 cm. (Take $\pi = \frac{22}{7}$.)

A *frustum* of a cone is a section cut off by a plane parallel to its base.

Let V be the vertex. PQ, SR are upper and lower diameters of the frustum. The line VNO is perpendicular to the base.

Let VQ $= x$ cm, then by similar triangles,[21]

[21] See *Geometry*, Teach Yourself Books, page 310, Theorem 58.

T.Y.A.—9

$$\frac{VQ}{VR} = \frac{NQ}{OR}$$

$$\therefore \frac{x}{x+25} = \frac{7}{12}, \text{ giving } 12x = 7x + 175$$

$$\therefore 5x = 175$$

$$\therefore x = 35.$$

$\therefore$ Curved surface of frustum

= Surface of cone VSR—Surface of cone VPQ

= $\pi RL - \pi r l$ cm^2,

where $R = 12$ cm, $L = (25 + 35)$ cm $= 60$ cm, $r = 7$ cm, and $l = x$ cm $= 35$ cm.

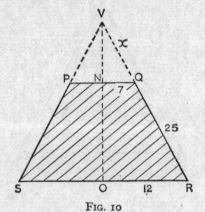

FIG. 10

$\therefore$ Curved surface $= \frac{22}{7} \times (12 \times 60 - 7 \times 35)$ cm^2
$= \frac{22}{7} \times 475$ cm^2
$\backsimeq$ **1490 cm^2**.

EXERCISE 3

1. Find the volumes of right circular cones, given that:
   (*a*) base radius is 5 cm, height is 7 cm,
   (*b*) slant height is 13 cm, base radius is 12 cm,
   (*c*) slant height is 5 m, vertical height 4·5 m.

2. Find the curved surface of right circular cones of:
   (*a*) base radius 4·5 cm, slant height 12·6 cm,
   (*b*) height 6 cm, slant height 8 cm,
   (*c*) height 4 m, base radius 2 m.

3. Find the *total* surface area of a solid cone of height 10 cm and base diameter 6 cm. (Note that the total surface includes the base.)

4. Find the number of cubic metres of air in a conical tent of height 3 m and diameter 2·4 m.

5. A heap of slag is in the form of a frustum of a cone, of upper and lower radii 4 m and 12 m respectively. The slant height of the heap is 10 m. Estimate the mass of the slag, if it is taken that 1 m³ is of mass 6 tonnes. (Give the answer in tonnes, to the nearest tonne.)

6. A solid steel cone of base radius 6 cm, height 12 cm, is dropped into a can partly filled with water. The can is a cylinder of radius 9 cm. If the cone is totally submerged, and no water flows out of the cylinder, find the rise in water-level, in centimetres (correct to 3 sig. figs.).

5. The Sphere

A *sphere* is traced out by a point moving so that its distance from a given point is constant. The definition is the same as for a circle except that the moving point is no longer restricted to a plane, but may move in space. This gives us a *hollow* sphere.

Alternatively we can think of a sphere as the *solid* generated by rotating a semicircle APB about its diameter AB.

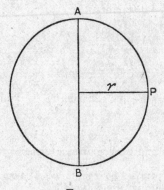

Fig. 11

If r is the radius of the sphere, its *volume* $= \frac{4}{3}\pi r^3$ and its *surface area* $= 4\pi r^2$.

These formulæ are obtained by calculus.

Ex. 10. Find the surface area and volume of a sphere of radius 2 cm.

Volume

$$= \tfrac{4}{3}\pi r^3 = \tfrac{4}{3} \times \tfrac{22}{7} \times 2^3 \text{ cm}^3$$

$$= \frac{4 \times 22 \times 8}{3 \times 7} \text{ cm}^3 = \mathbf{33 \cdot 5 \text{ cm}^3} \ (3 \text{ s.f.})$$

Surface area

$$= 4\pi r^2 = 4 \times \tfrac{22}{7} \times 2^2 \text{ cm}^2$$

$$= \frac{88 \times 4}{7} \text{ cm}^2 = \mathbf{50 \cdot 3 \text{ cm}^2} \ (3 \text{ s.f.}).$$

```
          88
           8×
      3)704
      7)234·66 ...
        33·52 ...


      7)352
        50·3
```

Ex. 11. The volume of a sphere is 10 cm³. Find its surface area.

If r cm is the radius

$$\tfrac{4}{3}\pi r^3 = 10$$
$$\therefore 4\pi r^3 = 30$$

so $r^3 = \dfrac{30}{4\pi}$, giving $r = \sqrt[3]{\dfrac{30}{4\pi}}$

(We can work out r here or proceed to the end before doing so.)

Now Surface area $= 4\pi r^2 \text{ cm}^2$

$$= 4\pi \left(\sqrt[3]{\frac{30}{4\pi}} \right)^2 \text{ cm}^2$$

$$= 4\pi \left(\sqrt[3]{\frac{7 \cdot 5}{\pi}} \right)^2 \text{ cm}^2$$

$$= 22 \cdot 45 \ \frown \ \mathbf{22 \cdot 5 \text{ cm}^2}.$$

No.	Log
7·5	0·8751
π	0·4971
—	0·3780
÷3	0·1260
×2	0·2520
4	0·6021
π	0·4971
+	**1·3512**

EXERCISE 4

1. Find the volumes and surface areas of spheres with the following radii:
   (a) 3·5 cm   (b) 3 cm   (c) 1 m   (d) 84·2 cm.
2. A sphere is of volume 4 m³. Find its radius.
3. The surface area of a sphere is 1000 cm². What is its radius?
4. Find the surface area of a sphere of volume 12 000 cm³.
5. A cube of metal, of side 8 cm, is melted down and formed into 1000 ball bearings. What is the radius of the bearings? (Answer in millimetres, correct to 3 significant figures.)
6. A boiler consists of a cylinder of radius 1·5 m and length 3·6 m with hemispherical ends. What is the capacity of the boiler in cubic metres? (Answer to 3 significant figures.)

7.

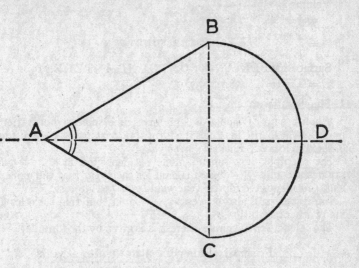

The figure illustrates a lamina which the line BC divides into a semi-circular part BDC and an equilateral triangle ABC (i.e. a triangle in which all sides are equal). If BC = 10 cm calculate the area of the lamina.

The lamina is now revolved about the line of symmetry AD. Calculate the volume of the complete solid formed, if the angle of rotation is 180°.

CHAPTER FOURTEEN

SIMPLE AND COMPOUND INTEREST

1. Simple Interest

When a sum of money is lent for a time some remuneration is made for the loan. This takes the form of *interest*, paid as a percentage of the sum loaned, called the *principal*.

For example, if a sum of money is deposited in a Post Office savings account, 5% interest is paid for each year that the money is lent. £200 deposited for one year would gain £10 interest.

Interest is paid also on deposits in banks, building societies and many other institutions.

The calculation of simple interest, I, is given by the formula

$$I = \frac{\text{Principal} \times \text{Rate per cent.} \times \text{Number of years}}{100}$$

or more shortly

$$I = \frac{P \times R \times N}{100} \quad \text{where} \quad \begin{cases} P = \text{Principal in £'s} \\ R = \text{Rate per cent.} \\ N = \text{Number of years.} \end{cases}$$

Ex. 1. Find the simple interest on £500 for 3 years at 4% per annum.[22]

Principal = £500, Rate per cent. = 4, Number of years = 3.

$$\therefore I = \frac{P \times R \times N}{100}$$

$$= £\frac{500 \times 4 \times 3}{100}$$

$$= £60.$$

When the time for which the loan is made is a number of days, the value of N is given by $\frac{\text{Number of days}}{365}$, and the calculation will nor-

[22] *Per annum* means for a year.

mally involve long multiplication and division and the use of decimals.

Ex. 2. Find the simple interest on £412 for 46 days at $3\frac{1}{4}\%$ per annum, to the nearest 1 p.

$P = £412, R = 3\frac{1}{4}, N = \frac{46}{365}.$

$$\therefore I = £\frac{412 \times 3\frac{1}{4} \times \frac{46}{365}}{100}$$

$$= \frac{\overset{103}{412} \times 13 \times 46}{100 \times 4 \times 365}$$

$$= £1\cdot6875$$

$$\frown £1\cdot69.$$

For interest on money, 4-figure logarithm tables are often insufficiently accurate. One can calculate, as here, or use (say) 7-figure tables as later in this chapter. An electronic calculator is better still. It usually has 8 working figures, and is very speedy in use, once the basic technique is mastered.

$$
\begin{array}{r}
103 \\
13 \times \\
\hline
1339 \\
46 \times \\
\hline
8034 \\
5356 \\
\end{array}
$$

$$365\overline{)615\cdot94}\,\,1\cdot687$$

$$
\begin{array}{r}
365 \\
\hline
250\cdot9 \\
219\cdot0 \\
\hline
31\cdot94 \\
29\cdot30 \\
\hline
2\cdot740 \\
2\cdot555 \\
\hline
185 \\
\end{array}
$$

2. Inverse Problems on Simple Interest

The formula $I = \dfrac{P \times R \times N}{100}$ involves four unknown quantities I, P, R, N, so if we are given any three we can find the fourth. To find any one of P, R, N it is convenient to rearrange the formula before using it.

We have

$$I = \frac{P \times R \times N}{100} \quad \bullet \quad \bullet \quad \bullet \quad (1)$$

$$\therefore \quad 100 \times I = P \times R \times N$$

$$\text{i.e. } P \times R \times N = 100 \times I \quad \bullet \quad \bullet \quad \bullet \quad (2)$$

Dividing (2) by $R \times N$ we have

$$P = \frac{100 \times I}{R \times N} \quad \bullet \quad \bullet \quad \bullet \quad (3)$$

Dividing (2) by $P \times N$ we have

$$R = \frac{100 \times I}{P \times N} \qquad \bullet \quad \bullet \quad \bullet \quad (4)$$

Dividing (2) by $P \times R$ we have

$$N = \frac{100 \times I}{P \times R} \qquad \bullet \quad \bullet \quad \bullet \quad (5)$$

The formulæ (3), (4) and (5) should be known, but it should be clearly understood that they are only variations of formula (1).

Ex. 3. What sum of money must be invested at $4\frac{1}{2}\%$ simple interest to give an interest of £90 in 5 years?

$R = 4\frac{1}{2}$, $I = £90$, $N = 5$.

$$\therefore P = \frac{100 \times I}{R \times N} = £\frac{100 \times 90}{5 \times 4\frac{1}{2}}$$

$$= £\frac{200 \times 90}{5 \times 9} = \textbf{£400.}$$

Ex. 4. For how long must £250 be invested at $7\frac{1}{2}\%$ per annum to give £75 simple interest?

$P = £250$, $I = £75$, $R = 7\frac{1}{2}$.

$$\therefore N = \frac{100 \times I}{P \times R} = \frac{100 \times 75}{250 \times 7\frac{1}{2}} = \textbf{4}$$

$\therefore$ The time is **4 years.**

EXERCISE 1

Find the simple interest due on the following investments:

	Principal	Rate per cent.	No. of years
1.	£400	6	3
2.	£285	$7\frac{1}{2}$	4
3.	£518	$3\frac{1}{2}$	5
4.	£6400	11	2.

Find the simple interest, to the nearest 1p, on:

	Principal	Rate per cent.	Time
5.	£274	$3\frac{1}{4}$	2 years
6.	£489	$5\frac{1}{2}$	81 days
7.	£3055	$8\frac{3}{4}$	175 days.

Find the missing items in the following table:

	Principal	Interest	Rate %	Time (years)
8.	x	£90	3	6
9.	£2400	£800	x	4
10.	x	£31·50	4½	3
11.	£1080	£432	5	x
12.	x	£638	7¼	5½

13. Jones pays £80 into the Post Office savings account and at the end of a year receives interest at 5%. He pays this interest into the same account. How much interest will he get at the end of the next year?

14. Smith wishes to raise a mortgage on his house. The interest charged is 9½% per annum. If he can only afford to pay £7·60 a week interest, how much money can he borrow? (It is to be assumed that the £7·60 is interest only and does not include any repayment of capital. The year is taken as 52 weeks.)

3. Compound Interest

The calculations made in simple interest are based on the assumption that the interest is paid in cash. It frequently happens, however, that the interest due is added to the principal at the end of each year, e.g. EXERCISE I, No. 13 above. In such a case the second-year principal is greater than that of the first year, for

2nd-year principal = 1st-year principal + 1st-year interest

The quantity on the right is called the *amount after one year*.

Thus, if we call 1st, 2nd, 3rd ... year principals P_1, P_2, P_3 ... and 1st, 2nd, 3rd ... year amounts A_1, A_2, A_3 ... and the corresponding interests I_1, I_2, I_3 ... then

$$P_2 = A_1 = P_1 + I_1$$
$$P_3 = A_2 = P_2 + I_2$$

Some examples will make this clear.

Ex. 5. Find the compound interest on £480 for 3 years at 5% per annum.

$$
\begin{aligned}
&\begin{cases} P_1 \\ I_1 \end{cases} \quad (5\% \text{ of } P_1) \quad &\begin{matrix} \pounds \\ 480 \\ \underline{24} \end{matrix} \\
&\begin{cases} P_2 = A_1 \\ I_2 \end{cases} \quad (5\% \text{ of } P_2) \quad &\begin{matrix} 504 \\ \underline{25\cdot2} \end{matrix} \\
&\begin{cases} P_3 = A_2 \\ I_3 \end{cases} \quad (5\% \text{ of } P_3) \quad &\begin{matrix} 529\cdot2 \\ \underline{26\cdot46} \end{matrix} \\
&A_3 \quad &\underline{\underline{\pounds555\cdot66}}
\end{aligned}
$$

Now A_3 is the amount for 3 years, so if we subtract from it the original principal P_1, we shall have the total interest added.

$$\begin{array}{lr} A_3 & \mathbf{555 \cdot 66} \\ P_1 & \mathbf{480 \cdot 00} \\ \text{Compound Interest} \longrightarrow & \mathbf{\pounds75 \cdot 66.} \end{array}$$

Notes

1. To find $x\%$ of a quantity move the decimal point two places to the left (thereby dividing by a hundred) mentally and multiply by x, e.g. 3% of 742

$$= 7 \cdot 42 \times 3 = 22 \cdot 26.$$

2. The normal number of decimal places required during working is *four* (unless the steps are exact as in Ex. 5 above, in which case it is sometimes unnecessary to put down as many).

Ex. 6. Find the amount of £238·83 for 2 years at 4% per annum.

	£	
P_1	$238 \cdot 83$	Note that we used 4
I_1 (4% of P_1)	$9 \cdot 5532$	decimal places to ensure
P_2	$248 \cdot 3832$	that the second decimal
I_2 (4% of P_2)	$9 \cdot 9353$	place was correct.
A_2	$\mathbf{\pounds258 \cdot 3185}$	

$\therefore$ Required amount is **£258·32.**

When the interest is a quantity like $3\frac{1}{2}\%$, we find 3% of the principal and $\frac{1}{2}\%$ separately and add the results. The $\frac{1}{2}\%$ can be found either as $\frac{1}{200}$ of the principal (involving moving the decimal point two places to the left mentally and dividing by two), or as $\frac{1}{6}$ of 3% (for $\frac{1}{2} = \frac{1}{6}$ of 3).

Ex. 7. Find by how much the compound interest exceeds the simple interest on £843 for 2 years at $3\frac{1}{2}\%$ per annum.

		£
P_1		$843 \cdot 00$
I_1	$\begin{cases} 3\% \\ \frac{1}{2}\% \end{cases}$	$25 \cdot 29$ $4 \cdot 215$
P_2		$872 \cdot 505$
I_2	$\begin{cases} 3\% \\ \frac{1}{2}\% \end{cases}$	$26 \cdot 1752$ $4 \cdot 3625$
A_2 (amount for 2 years)		$903 \cdot 0427$
less P_1		$843 \cdot 0000$
Compound Interest		$\mathbf{\pounds60 \cdot 0427}$

$$\text{Simple interest} = \frac{843 \times 2 \times 3\frac{1}{2}}{100} \quad \Big| \quad \begin{array}{l} 843 \times 7 \\ = 5901 \end{array}$$

$$= \pounds 59\cdot 01$$

$\therefore$ Compound interest — Simple interest

$$= \pounds(60\cdot 0427 - 59\cdot 01)$$
$$= \pounds 1\cdot 0327$$
$$\frown \pounds 1\cdot 03.$$

It will be observed that the method used above is *practice*, and on occasion it can be tedious. In section 4 below an alternative method of great practical importance is discussed. When the number of years is great, practice methods are useless and the use of tables or of an electronic calculator is necessary.

Errors which arise in calculation by the method of the present section (i.e. section 3) can sometimes be spotted by noting that the successive values of the interest I increase fairly steadily.

EXERCISE 2

Find the compound interest on the following loans:

	Principal	Rate per cent. p.a.[23]	No. of Years
1.	£640	5	2
2.	£1750	4	3
3.	£184	8	4
4.	£1938	$3\frac{1}{2}$	2
5.	£643	$4\frac{1}{2}$	3.

6. Find the compound interest on £720 for 2 years at 7% per annum, payable half-yearly. (In this case, take half the interest rate, $3\frac{1}{2}$%, and add it in every half-year, i.e. 4 times in all.)

7. Find by how much the compound interest exceeds the simple interest on:
   (a) £400 for 2 years at 3% per annum
   (b) £2750 for 3 years at $5\frac{1}{2}$% per annum.

4. Compound Interest Formula

We saw that the second-year principal was the same as the first-year amount, i.e. $P_2 = A_1$, but

$$A_1 = P_1 + P_1 \times \frac{r}{100} \text{ (for it is } P_1 \text{ plus } r\% \text{ of } P_1\text{).}$$

$$\therefore P_2 = A_1 = P_1\left(1 + \frac{r}{100}\right)$$

[23] per annum (i.e. each year)

Also $A_2 = P_2\left(1 + \dfrac{r}{100}\right)$

$\qquad = P_1\left(1 + \dfrac{r}{100}\right)\left(1 + \dfrac{r}{100}\right)$

$\qquad = P_1\left(1 + \dfrac{r}{100}\right)^2$

Proceeding in this way we see that the amount for n years, A_n, is given by

$$A_n = P\left(1 + \frac{r}{100}\right)^n \qquad \bullet \quad \bullet \quad \bullet \quad \bullet \quad (1)$$

where P is the original principal and r is the rate per cent. per annum.

If we put $R = 1 + \dfrac{r}{100}$ in equation (1) we get the alternative formula

$$A = P.R^n. \qquad \bullet \quad \bullet \quad \bullet \quad \bullet \quad (2)$$

The quantity R is called the amount of £1 for 1 year, for if we put $n = 1$, $P = £1$ in equation (1) we get $A = £\left(1 + \dfrac{r}{100}\right)$.

The formula (2) is of considerable value in calculating A, especially when n is a large number of years. The value of R^n can be obtained from a set of tables such as that given below.

AMOUNT OF £1 AT COMPOUND INTEREST

Years	2½%	3%	3½%	4%	5%
1	1·025 000 0	1·030 000 0	1·035 000 0	1·040 000 0	1·050 000 0
2	1·050 625 0	1·060 900 0	1·071 225 0	1·081 600 0	1·102 500 0
3	1·076 890 6	1·092 727 0	1·108 717 9	1·124 864 0	1·157 625 0
4	1·103 812 9	1·125 508 8	1·147 523 0	1·169 858 6	1·215 506 2
5	1·131 408 2	1·159 274 1	1·187 686 3	1·216 652 9	1·276 281 6
10	1·280 084 5	1·343 916 4	1·410 598 8	1·480 244 3	1·628 894 6
20	1·638 616 4	1·806 111 2	1·989 788 9	2·191 123 1	2·653 297 7
30	2·097 567 6	2·427 262 5	2·806 793 7	3·243 397 5	4·321 942 4

Ex. 8. Find the amount of £650 for 10 years at $3\frac{1}{2}\%$.

Reading off the value of R^n in the row opposite 10 years and in the column under $3\frac{1}{2}\%$, we get

$$R^n = 1\cdot410\ 598\ 8.$$

∴ Required amount A

= £650 × 1·410 598 8

≏ **£916·89.**

$$\begin{array}{r} 14·105\ 988 \\ 65\ \times \\ \hline 70·529\ 940 \\ 846·359\ 28 \\ \hline \mathbf{916}·889\ 22 \end{array}$$

It will be observed that, although 7-figure tables are used, these are not of too great an accuracy when multiplying by sums of money which run into hundreds or thousands of pounds.

5. Compound Interest by Logarithms

Consider again the equation $A = PR^n$

$$\left(\text{where } R = 1 + \frac{r}{100}\right).$$

If we take logarithms we have [24]

$$\log A = \log P + n \log R \qquad . \qquad . \qquad . \quad (3)$$

and this gives us a method alternative to that shown in section 4 above. It is particularly useful, if a table of amounts of £1 is not available.

Ex. 9. Find the compound interest on £749 for 8 years at $4\frac{1}{2}\%$.

We have $R = 1 + \dfrac{4\frac{1}{2}}{100} = 1·045$, $n = 8$, $P = £749$.

	No.	Log
	1·045	0·0191
∴ A = PR⁸		
= £749 × (1·045)⁸	×8	0·1528
= **£1065.**	749	2·8745
	+	**3·0273**

$\therefore A = PR^8$
$= £749 \times (1·045)^8$
$= \mathbf{£1065.}$

This answer is an approximation, as only 4-figure tables are used. As before, for accuracy, 7-figure tables are desirable.

Inverse problems on compound interest can readily be solved using logarithms.

Ex. 10. £500 is lent for 3 years, and it is understood that, including interest, £600 is to be repaid at the end of that time. What percentage compound interest does this represent, correct to 1 decimal place?

[24] For $\log A = \log PR^n = \log P + \log R^n = \log P + n \log R.$

Here we have A = £600, P = £500, $n = 3$.

	No.	Log
	1·2	0·0792
	$\div 3$	0·0264

$\therefore$ $\quad 600 = 500 \, R^3$

so $R^3 = \frac{6}{5}$, i.e. $R = \sqrt[3]{1·2}$

$\therefore$ $\quad R = 1·063$

i.e. $1 + \dfrac{r}{100} = 1 + ·063$

$\therefore$ $\quad$ **r = 6·3**

i.e. **percentage interest is 6·3.**

6. Repayment of Loans by Instalments

A common method of repayment of a loan is by instalments once a year. In such a case, for each year the interest on the outstanding principal is added, and the amount of the annual repayment is then subtracted. An example will make this clear.

Ex. 11. Brown buys a car for £2850 and borrows the money at 7% per annum compound interest. He agrees to make annual repayments of £800, the first to be paid twelve months after the loan is issued. How much does he still owe immediately after making the third repayment?

	£
1st-year principal	2850·00
Add 7% interest	199·50
	3049·50
Less repayment	800·00
2nd-year principal	2249·50
Add 7% interest	157·47
	2406·97
Less repayment	800·00
3rd-year principal	1606·97
Add 7% interest	112·49
	1719·46
Less repayment	800·00
Outstanding debt	**919·46**

The amount still owing at that date = **£919·46.**

Using the Compound Interest Table in section 4 above, find to the nearest 1p:

1. The amount of £580 for 4 years at 3%
2. The amount of £1200 for 10 years at 5%

3. The amount of £685 for 20 years at 2½%

4. The interest on £2800 for 5 years at 4%.

Using 4-figure logarithms, find as accurately as you can:

5. The amount of £820 for 6 years at 4½%

6. The amount of £2040 for 11 years at 3%

7. The compound interest on £722 for 8 years at 7½%

8. The compound interest on £28 000 for 6 years at 4%.

9. At what rate per cent., to the nearest 0·1%, is interest charged if a repayment of £750 is to be made at the end of 5 years for a loan of £600?

10. In how many years would a sum of money double itself if invested at (*a*) 4% compound interest, (*b*) 8% compound interest?

11. The compound interest paid on £400 loaned for two years exceeds the simple interest on the same amount for the same time at the same rate per cent., by £4. Find the ráte per cent. per annum.

It is recommended (see p. 196) that the student, if not already in possession of one, might do well to acquire an electronic calculator. Such machines can save much time in evaluating results, but they do, however, vary greatly in their capacity to solve problems and, of course, in their corresponding cost.

For general purposes, a simple machine having at least the following functions should be adequate: $+ - \times \div$, floating decimal, square and square root, % and full-access memory. The solution of many of the problems in this and in other chapters would be easier with such a calculator. However, certain questions could not be solved without the use of a more elaborate *scientific* calculator (e.g. p. 148, Ex. 11 and p. 157, Ex. 10) involving cube roots which need logarithms for their solution).

CHAPTER FIFTEEN

RATES, TAXES AND INSURANCE

1. Rates and Taxes

It is not within the scope of this book to give a detailed account of national expenditure or that of local government, but as the costs entailed are borne by the public, everyone should understand the arithmetic involved.

Rates are paid to local authorities, such as Borough or Urban District Councils, for public services. Some of these services are provided by the local council, others are controlled by the County Council, who receive a proportion of the rates paid to the local council. For example, an Urban District Council will deal with housing, parks and cemeteries and public libraries. A County Council will deal with education and fire service, but consulation between the two authorities is necessary for many items.

National expenditure is largely met by the payment of *taxes*. These take many forms including Income Tax, Value Added Tax, Capital Gains Tax and Capital Transfer Tax. The joint effect of Income Tax and Purchase Tax alone is so heavy that frequently, if not usually, they will exceed 50% of the actual *cost* of an article purchased.

2. Rates

All property, such as houses, shops, factories and business premises, in an area is *assessed*. Each property is given an *assessment* or *rateable value*. The calculation of the assessment is rather involved and depends to some extent on a hypothetical net income which could be received if the property were let. Contributory factors in assessing the rateable value of a house are its locality, floor area, size of garden, whether or not there is a garage, and so on. Once the assessment has been made (for an ordinary house the figure might well be £240) it will remain constant for a number of years. The *rates* paid, however, do not. They will vary from year to year and they depend on the annual expenditure of the local authorities. In this connection a simple proportion sum is involved.

Suppose a Borough Council wishes to raise £2 400 000 to cover its expenses, and let us assume that the rateable value of the Borough is £4 000 000. Then the rate levied is $\frac{2\ 400\ 000}{4\ 000\ 000}$ of the rateable value, i.e. out of each £1 of rateable value, the council needs

$$£\frac{2\,400\,000}{4\,000\,000} = \frac{3}{5} \times 100p = 60p.$$

We say that **a rate of 60p in the £** is levied.

Taking our house, mentioned above, as having a rateable value of £240 the actual amount (rate) to be paid is

$$240 \times 60p = 14\,400p = £144.$$

Actual calculations do not work out as conveniently as this example and they are made using decimals of a penny, as will be seen from the extracts from the Rate Demand Note illustrated below.

Ex. 1. The total rateable value of a Borough is £4 390 000. The estimated expenditure for a certain year £3 160 000. Find the rate in the £ to be levied to the nearest 1p.

$$\text{Rate to be levied} = \frac{3\,160\,000}{4\,390\,000} \times 100p$$

$$= 72p \text{ (to the nearest 1p).}$$

Ex. 2. The public lighting in a certain district of rateable value £2 370 000 required a levy of 3·40p in the £. How much money was needed?

$$\text{Amount required} = 2\,370\,000 \times 3·40p$$

$$= £80\,580.$$

EXTRACT FROM A SPECIMEN
RATE DEMAND NOTE

	Rate Levied by	
	U.D.C.	County Council
	new pence	new pence
Housing	1·05	—
Education	—	82·76
Public Health etc.	6·71	8·43
Police etc.	—	11·58
Miscellaneous	14·36	15·60
Total for all purposes →	22·12	118·37
Deduct in respect of		
(a) Housing Subsidies	1·05	—
(b) Government Grants etc.	—	71·44
Total deductions →	1·05	71·44
Rate in the pound payable →	21·07	46·93

Rate in pound payable by ratepayer is

$$(21{\cdot}07 + 46{\cdot}93)\text{p} = \textbf{68p.}$$

EXERCISE I

1. The rate charged is 64p in the £. How much will Brown have to pay if his house has a rateable value of £156?

2. If the rateable value of a district is £1 500 000 the rate charged is 76p in the £. If the rateable value is increased to £2 000 000 in the same district, what rate in the £ will give the same income to the council?

3. What rate in the £ is charged if £3 465 000 is required by a Borough in which the rateable value is £5 293 000? (Give the answer to 0·01p.)

4. The rateable value of an urban district is £4 864 000. How much will a levy of 3·54p in the £ raise?

3. Water Rate

Until recently, the Water Rate was charged through the office of the local council. Now, however, it seems that the Regional Water Boards are bringing into use their own offices for this purpose.

EXAMPLE OF WATER RATE DEMAND NOTE

Water Supply, Sewerage and Environmental Services

Rateable Value Chargeable	Water Supply Charge 8·28%	Sewerage Charge 10·20%	Fixed Charge	Credit/ Debit Brought Forward	TOTAL AMOUNT DUE £
£250	20·70	25·50	3·75	—	49·95

4. Income Tax

Income Tax is the largest single source of revenue for the Exchequer. This, together with Capital Gains Tax, Capital Transfer Tax, Stamp Duties and Vehicle Licence are the main forms of direct taxation, while Value Added Tax and Customs and Excise Duties, including those on alcohol and tobacco, constitute the main body of indirect taxes.

Everyone who has a source of income must declare it to his local Inspector of Taxes, so that the tax to be paid on his income can be calculated. The calculation of Income Tax varies almost every year, so examples must perforce be based on the figures used in 1978–9.

Although the numbers involved are subject to substantial variation, the method can be applied at any time, and is best illustrated by a fairly comprehensive example.

Ex. 3. A married man with two children earns £5750 in the Income Tax Year 1978–79. He pays a superannuation contribution of 6% of his salary. His personal allowance as a married man is £1295 and the allowances for his children are as follows: for Louise, born in 1963, £231, and for George, born in 1969, £170. There is an endowment assurance policy on which the premium is £240 p.a. and for which 17% of the premium is allowed. The standard rate of tax applicable to the residue of his income is at 34% (i.e. at 34p in the £). Calculate the Income Tax due.

The order of procedure is shown below.

	£	£
EMPLOYMENT etc.		5750
less Superannuation (6%)	345	
Allowable Expenses	—	
	345	345
		5405
Building Society Interest	—	
Personal Allowance (married man) . .	1295	
Wife's Earned-Income Allowance . . .	—	
Additional Personal	—	
Children	401	
Dependent Relatives	—	
Life Insurance (17% of premium allowed) .	41	
TOTAL ALLOWANCES	1737	
less Allowances made elsewhere . .	—	
	1737	1737
Net Amount chargeable to tax . . .		3668
TAX CHARGEABLE:		£
Basic Rate (34%) on £3668 . . .		1247 12
Higher Rate ( ) on . . .		— —
TAX ASSESSED FOR 1978		**1247 12**

Thus the taxpayer could expect to have the sum of £103·92 a month (one-twelfth of £1247·12) deducted from his salary under P.A.Y.E. (Pay As You Earn).

The above example is laid out in a manner similar to that of Form P70C (Part I). Some items on the original have been omitted as they are easily understood, and they are not necessary for the calculation required in this problem. This form is sent to all salary- or wage-earners each year and it is extremely important that it should be clearly understood by everyone.

One or two points in the above calculation require elucidation. (1) The Superannuation is calculated on the gross (total) earned income. (2) The allowance for the children is £231 for Louise *plus* £170 for George, giving £401 in all. (3) The Life Insurance (wrongly worded on the form—it should read Life Assurance) is $\frac{17}{100} \times £240 \frown £41$, to the nearest £1.

The assessment of the amount due from any individual tax-payer is a specialised problem. There are many factors to take into account and, in these days when there may be two or more budgets a year, the Inspectors of Taxes are themselves hard pressed to cope with the varying regulations. It is hoped, however, that this short section on taxation will enable those readers who, like the present author, approach their assessment forms with sinking hearts, at least to follow the calculations involved.

The form P70C should be kept permanently, as it is by no means unusual for tax adjustments to be made for income received several years earlier.

The above calculations refer to Schedule E income, but there are various other schedules on which tax may be due. For further information, the reader is referred to books specialising on the subject.

5. Insurance

Insurance in modern times is a complicated matter, and this section can only serve as a brief introduction to the subject. An excellent comprehensive treatment is given in Mr Cockerell's book, *Insurance*, in the Teach Yourself series, a companion volume to this book.

Briefly, insurance can be classified into two sections:

(*a*) Life Assurance
(*b*) Property Insurance.

(*a*) Life Assurance

The following table shows the division of this section.

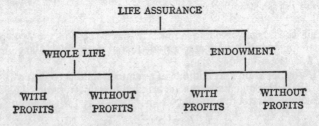

A *whole life* policy is one in which the assured person cannot himself benefit. His dependants receive a sum of money in the event of his death.

An *endowment* policy is one taken out for a specified number of years (say 10, 15, 20, 25 or 30 years). If the policy-holder survives the date of maturity of the policy he can himself receive the sum assured. If he should die before this time his dependants will receive the money at the time of his death.

A *with-profits* policy shares in the profits of the company, and every few years (say, three or five) a bonus is declared and is added to the capital sum assured. On the other hand, a *without-profits* policy does not participate in this way. The sum assured remains constant.

The following table for 'Endowment Assurances—With Profits' will illustrate the principles. The sum assured is payable at the end of the period of years specified or at previous death.

ANNUAL PREMIUM FOR EACH £100 ASSURED

Age Next Birthday	Number of Years				
	10	15	20	25	30
	£	£	£	£	£
26	10·94	7·34	5·51	4·42	3·68
27	10·95	7·35	5·51	4·43	3·69
28	10·95	7·35	5·52	4·44	3·70
29	10·96	7·36	5·53	4·45	3·72

(The table above does not necessarily represent any company's current quotations, but it illustrates the kind of charge made.)

Ex. 4. Thomas wishes to take out an endowment policy with profits for 25 years. The capital sum assured is £800. Thomas is 27 years old and in good health. How much is his annual premium, if the basic tax rate is 35%?

From the table above, reading off in the row opposite 28 (note that it says age *next* birthday) under the column for 25 years, we find that the premium is £4·44 per £100.

$\therefore$ Annual premium to be paid = £4·44 × 8
= **£35·52.**

It is worth bearing in mind that the actual cost each year is appreciably less. If Thomas pays Income Tax, the net cost would be obtained after allowing for a scale of tax relief allowance.

In general, and with certain limitations, premiums paid for life assurance are subject to Income Tax relief at the appropriate rate on (a) the amount of the premiums paid in the year of assessment, if less than £10, (b) £10, if the premiums paid are between £10 and £25, (c) the whole premium at half the basic rate of tax, if the premium exceeds £25.

Thus, Thomas' net outgoing would be:

		£
Annual premium		35·52
less 17·5% of £35·52		6·22
Net outgoing		£29·30

As from 6th April, 1979, eligible life assurance premiums are paid *net* (as above) to the Life Assurance Office; this office will itself make good to the inspector of taxes the 'underpayment' by the policy-holder.

In order to reduce expenses of collection, it has been agreed *for this purpose* that an assumed basic tax rate of 35p in the £ shall apply from that changeover date (6.4.79).

A comparison of premiums is interesting for, say, present age 27 years.

1. Endowment with profits (25 years)	£4·44 ⎞	
2. Endowment without profits (25 years)	£3·37 ⎟ per	
3. Whole life with profits[25]	£2·42 ⎟ £100	
4. Whole life without profits[25]	£1·62 ⎠	

There is a big difference in premium for each class of assurance, but there is also a big difference in cover. There is little doubt that, if the higher cost can be met, then endowment assurance with profits is by far the best for most purposes.

(b) Property Insurance

The most important forms of insurance for the average man are:

    (1) House insurance (if owner of the house);
    (2) Contents insurance;
    (3) Car insurance.

Items (1) and (2) above can be considered together but item (3) is always dealt with separately. Whereas the costs of (1) and (2) are directly related to the amounts insured, the premium for (3) is related to the make and model of the vehicle, and also the driver's record, etc. The premium depends on the number of claims made during a year. Those made for the theft of property or for fire damage are few

[25] Premiums payable for life.

compared with the enormous number of costly claims made for motor traffic accidents.

At the time of writing, house insurance can be effected at a satisfactory standard for a premium of about 14p for £100 cover. Contents cost a little more, perhaps 30p per £100; there is some dependence here on whether there are specific items to be insured separately.

Ex. 5. Harris bought a house worth £17 000 and valued the contents at £6000. How much a year would his premium be for both, if the house was insured at 14p per £100 and the contents at 30p per £100?

Premium for house 170 × 14p = £23·80
Premium for contents 60 × 30p = £18·00

Annual premium altogether   **£41·80**

It is sad, but Income Tax rebate can be claimed for few forms of insurance other than life (including endowment policies), superannuation or pension insurance. Thus, no relief can be claimed for Harris's premium.

The calculations for car insurance are different for almost every car and driver, so no attempt will be made to deal with them here. It is best to get quotations from different companies as there are appreciable variations in vehicle insurance, but it should be borne carefully in mind that the cheapest forms of insurance are very far from being the best. For the owner of a car, a comprehensive policy with a good insurance company or with Lloyd's is a valuable safeguard. Incidentally, motor insurance (personal injury/third party) is compulsory under law.

EXERCISE 2

Lay out the following (Questions 1, 2, 4) in the form of the example on Income Tax given in Section 4 above. Take the rate of tax as the same.

1. A single man earns £4800 in the tax year 1978–79. He pays 6% superannuation contribution. His personal allowance as a single man is £845. How much tax does he pay?

2. A man earning £6250 has a wife (his married man's allowance being £1295) and one child (the allowance for the latter being £196 because of the child's age). The man does not pay superannuation, but he has a life assurance policy on which the annual premium is £300. Find the Income Tax paid.

3. Using the table of premiums shown on p. 165, find the *net* annual cost for an endowment assurance policy (with profits) on which the capital sum is £2400, for a man aged 28½, if the policy is to run for 20 years.

4. Jones earns £3400 p.a. He pays 5% superannuation contribution. His married man's allowance is £1295. He has three children, Hermione, for whom the allowance is £261, Alfred, for whom it is £205, and Eleanor, for whom it is £170. Jones has an endowment assurance policy on which the annual premium is £200. Calculate the tax he pays.

5. Brown, who is 27 years of age, wishes to take out an endowment policy with profits. The maximum *net* premium he can pay each year is £167. If the policy is to run for 15 years, find the maximum capital sum which he can afford to take up.

(c) Other Classes of Insurance

In addition to the above, very widely used, forms of assurance and insurance, the following are also worthy of brief mention:

(a) Assurance protection in the *purchase* of a house
(b) School fees
(c) Pension arrangements
(d) Creditor assurance
(e) Partnership and Co-director arrangements
(f) Trustee reassurance.

These are more specialised topics, and the reader is advised to seek information on them elsewhere.

INVESTMENTS AND THE
STOCK EXCHANGE

1. What are Stocks and Shares?

Stocks and shares are holdings, i.e. units of part-ownership, in a company or organisation. Anyone who buys such investments becomes a part-owner of the concern, no matter how few shares he may purchase. It does not necessarily follow that he has a say in the running of the businesss.

In the case of public companies, equity shareholders are usually invited to attend at Annual General Meetings, where they may have an opportunity of expressing ideas, and where they may vote. The power of their vote is normally proportional to the number of shares held. Shareholders are also able to elect directors or to remove them from office.

The issue of *shares* in companies is normally a permanent matter. So long as the company exists, the shares continue in being. There are certain *loans* raised in developing new premises or activities which are of a temporary nature. They often take the form of *mortgage debentures* on the property of companies or *unsecured loan stock*, or they may be issued in the form of *redeemable preference shares* (*q.v.*). Loan notes are also sometimes used.

Government loans and those made to local Public Boards are not quite the same. They are normally of two kinds:

(1) A loan may be for a specific period, by the end of which time it will have been redeemed. For example, $8\frac{1}{4}$% Treasury Stock (1987–90) could be redeemed[26] as early as 1987, but *must* have been redeemed by 1990.

(2) The loan may be for an indefinite period, e.g. 4% Consolidated Stock (called Consols for short) is dated *in or after* 1957. This means that the issuing authority, in this case the Government, *can* redeem the stock at any time in 1957 or later, or need not do so at all. So long as the stock is held, interest is paid at 4% per annum.

---

[26] A stock is redeemed by an issuing authority when the holders are paid cash for their stock. The loan then ceases. Government and Local Authority stock is usually repaid at *par*, i.e. £100 is paid for £100 stock.

2. The Stock Exchanges

The Stock Exchanges are semi-public institutions where the purchases and sales of stocks and shares are carried out. There are exchanges at Manchester, Birmingham and other provincial cities, as well as in London, but there seem to have been some suggestions about the creation of a single United Stock Exchange.

A seller who wishes to dispose of a security can, via the stock exchanges, quickly and easily be introduced to a buyer. A company which seeks financial aid can arrange for a loan in the form, say, of a debenture issue. Stock exchange business is transacted by members, the public not being admitted for this purpose.

Suppose Jones wishes to sell 200 Ordinary Shares of I.C.I. (Imperial Chemical Industries). He telephones or visits his broker (or he may instruct his bank to contact a broker) and tells him to sell the security. If he knows the market value he may also state a minimum price,[27] say 270p, which he wishes to obtain. His *stockbroker*, in his daily visit to the Stock Exchange, contacts a *stockjobber* who is willing to take the shares on his book at an agreed figure. To get the highest price for his client a broker may contact several jobbers before he is satisfied that he has done as well as possible. Jobber No. 1 may quote 272p–274½p, meaning that he will *give* 272p for I.C.I. but *charges* 274½p for them. Jobber No. 2 may quote 273–275½p. Thus, Jones's broker would get a better price from Jobber No. 2, and, as this would be 3p more than Jones's minimum figure, a deal might well be made. The jobber does not keep the 200 shares. He waits until another broker comes to him from another client, say Brown, who wishes to buy I.C.I. and, if jobber and broker agree the price, the shares are bought by the broker on behalf of Brown.

It will be noticed that a *stockjobber* is mentioned above. He does not deal directly with the public. He is a man of considerable financial means who holds large quantities of stocks and shares between transactions.

The full story[28] of Jones's sale would read as follows:

JONES→BROKER A→JOBBER→BROKER B→BROWN

The broker charges a small commission on each share for his work on behalf of his client. The jobber makes his income on the difference in price between buying and selling, e.g. the jobber who quoted, say, 273p–275½p would make a profit of 2½p on each share if he obtained the price asked. On 200 shares this may seem a lot, considering the number of transactions he will carry out in one day, but it should be

[27] For each share, quoted in new pence.
[28] See paragraph 7 for the documents involved.

borne in mind that he holds large quantities of securities on his hands at any one time, and should some of them fall in value during the day, a very frequent event, he may stand to lose substantial sums of money from time to time.

The *broker* does not hold the stock himself, but acts as an agent on behalf of his client. Should a client default, the broker will have to honour the bargain. There exists a compensation fund to protect the investor. For this reason, the broker takes care to check the *bona fides* of his clientele.

3. Stocks and Shares

The term 'stock' is usually restricted to official issues such as British Funds (e.g. 3.5% War Stock), together with various classes of loan stock. The War Stock price, quoted in *The Financial Times*, the *Daily Telegraph* and other leading newspapers, might be found to be 34. This means that the average price of £100 *Stock* (i.e. stock on the *nominal value*[29] of £100) stands at *about* £34 on that day. The £100 stock is *not* worth £100 in cash, although it *may* have been issued at that price many years ago. The *market value*,[30] £34, is determined by the income provided, calculated as a percentage of the cash outlay. The safety of the capital outlay is also an important consideration. When a high percentage return is expected, the market price falls, and *vice versa*.

When a company is formed, people are asked to subscribe towards it by taking 'shares'. The shares are usually for small amounts, such as £1, 50p, 25p, 10p and 5p, although shares can be found in many shapes and sizes. The shares are rarely worth the same as their nominal value. The price depends not only on yield, as with some stocks, but on potential capital appreciation.

Consider, again, the purchase of £100 of 3.5% War Stock at 34. The cost if £34, but the *dividend* of 3.5% is based on the nominal value of the certificate, which is 100, so we have

COST	CERTIFICATE		INCOME
£34 CASH $\longrightarrow$	£100 of 3.5% War Stock	$\longrightarrow$	£3.50 per annum

The yield, expressed as a percentage of outlay, is much more than $3\frac{1}{2}\%$, for £34 brings an annual interest of £3.50.

[29] Definitions are given in Section 4, p. 172. The nominal value is that written on the stock or share certificate. Quotations for stocks of this nature are given in £'s, not in pence.

[30] Definitions are given in Para. 4.

$$\therefore \text{ Percentage yield} = \frac{3 \cdot 5}{34} \times 100 \mathbin{\frown} \mathbf{10 \cdot 29.}$$

4. Definitions of certain terms used in Stock Exchange procedure

(a) The *nominal value* of a security is the amount of stock or shares held. It is written on the stock or share certificate. It does *not* vary from day to day.

(b) The *market value* of a security is the value, in money, of the holding, and this may fluctuate from day to day.

(c) A *quotation* is the price to be paid for a share or unit of stock. This also varies. Note that:

$$\text{market value} = \text{quotation} \times \text{number of shares.}$$

(d) A stock is said to stand:

(1) *at par* if it is quoted at its nominal value exactly, e.g. £100 Stock standing at £100 cash, i.e. quoted at 100;

(2) *at a premium* if it is quoted above par, e.g. a £1 share quoted at 178p stands at a premium of 78p;

(3) *at a discount* if it is quoted below par, e.g. Consols 4% standing at £36 would be at a discount of 64. It would cost £36 for £100 Consols.

(e) The *dividend* received is the amount of cash paid as interest on a holding. It may be expressed as a percentage of the stock, as in the Consols above, or it may be given as a sum of money, such as 3p dividend on each 25p ordinary share. The latter would be

$$\tfrac{3}{25} \times 100\%, \text{ i.e. } 12\% \text{ dividend.}$$

(f) The *yield* is the dividend expressed as a percentage of the *cost price*. For example, if the 25p shares which paid 3p dividend in (d) above cost 48p each, the yield would be less than the 12% dividend. It would be

$$\frac{25\mathrm{p}}{48\mathrm{p}} \times 12\% = \frac{25 \times 12}{48}\% = 6 \cdot 25\%.$$

5. Methods of Calculation

The calculations involved in the purchase and sale of stock are best illustrated by some examples. Those in the first group below will avoid the complications of brokerage and stamp duty, but later some explanation of these expenses is given.

Ex. 1. Find how much $7\frac{3}{4}\%$ Treasury Stock (2012–15) at 75 can be bought for £600. Calculate also the annual income.

(The stock stands at a discount so the quantity of stock bought will exceed the cash value.)

$$£75 \text{ cash buys } £100 \text{ stock}$$

$$\therefore \text{ Amount of stock bought} = £\frac{100}{75} \times 600$$

$$= \textbf{£800 stock.}$$

$$\text{Annual income} = £\frac{7 \cdot 75}{100} \times 800$$

$$= \textbf{£62.}$$

Ex. 2. Smith invested £480 in Commercial Union 25p Ordinary Shares quoted at 150p. The annual dividend paid was $7\frac{1}{2}$p net per share. Find (a) the number of shares bought, (b) the actual income received and (c) the yield per cent. on outlay.

The student will notice that the shares are priced very highly, but the dividend is also high compared with the nominal value of the shares, so that a moderate yield is to be anticipated.

The number of shares purchased is quite independent of their nominal value. They cost 150p, i.e. £1·50, each.

$$\therefore \text{ Number of shares purchased} = \frac{480}{1 \cdot 5}$$

$$= \textbf{320.}$$

(Their total *nominal value* is only $320 \times 25p = £80.$)

$$\text{The dividend received is } 7\frac{1}{2}\text{p per share}$$
$$\text{i.e. altogether it is } 7 \cdot 5 \times 320p = \textbf{£24}$$

and this is the annual income.

$$\text{The yield per cent. on outlay} = \frac{£24}{£480} \times 100$$

$$= \textbf{5\%}$$

This example is interesting in that it illustrates the flaw in certain popular fallacies concerning dividends. There are many who believe that the payment by a company of 50% dividend implies enormous profits to the shareholders. This is not so, for the above example shows that whereas the dividend of $7\frac{1}{2}$p per 25p share is exactly 30%, the yield on capital gained by a shareholder is only 5%, e.g. far less than he would get if he bought National Savings Certificates.

A company pays dividends in accordance with its profits. The company may grow enormously over many years of well-managed business, and may employ more and more people and have greatly increased premises. It does not usually increase its share capital (i.e. the number of shares issued) to anything like the same extent, so the value of the individual shares will rise. The percentage interest received depends on *yield* based on the cost of the share, i.e. the market value, not on the dividend, which is based on the nominal value.

Ex. 3. Thompson invested £1344 in Shell Transport 25p shares, paying 60% net dividend. He received an income of £30. At what price did the shares stand when he bought them?

The dividend from one share $= \frac{60}{100} \times 25$p

$$= 15\text{p}.$$

∴ If x is the number of shares, the yield is $15x$ pence, but we know that this is £42 = 4200 pence

$$\therefore 15x = 4200$$
$$\text{so} \quad x = 280$$

and this is the number of shares bought.

∴ 280 shares cost £1344

i.e. 1 share costs £$\frac{1344}{280}$ = £4·80 = **480p.**

Ex. 4. Messrs Atkinson & Wright Ltd. hold £10 000 3% Exchequer Bonds. They sell their investment at 77 and reinvest the proceeds in Funding $6\frac{1}{2}$% Stock at 60. Find the change in annual income.

This type of problem should be divided mentally into stages.

(1) *Original Income* $= \dfrac{3}{100} \times$ £10 000 = £300

(2) Amount of money realised on sale
$$= \text{£10 000} \times \tfrac{77}{100}$$

(3) Amount of stock purchased $= \tfrac{100}{60} \times$ sum invested
$$= \text{£10 000} \times \tfrac{77}{100} \times \tfrac{100}{60}$$

(4) *New income* $= \dfrac{6\cdot5}{100} \times$ amount of *new* stock held

$$= \text{£10 000} \times \frac{77}{100} \times \frac{100}{60} \times \frac{65}{1000}$$

$$= \text{£834·17.}$$

∴ There is an increase in income of

$$\text{£}(834\cdot17 - 300) = \text{£534·17.}$$

Exercise 1

1. Find how much 3·5% War Stock quoted at 33 can be bought for £2640. Also find the annual interest received.

2. Brown invests £315 in 2½% Consols at 21. What is the yield per cent. and what is his annual income?

3. Which gives the higher yield: (a) 5·5% Funding Stock at 76 or (b) 3·5% Funding Stock at 48? (Assume that £100 has been invested in each security, and compare the results.)

4. What is the cost of 250 I.C.I. £1 units of Ordinary Stock at 364p each? If they pay 16½p dividend, what is the annual income from them and what is the yield per cent.?

5. Jones bought 240 New Organisation 25p Ordinary Shares for £1104. At what price were they quoted? At this price the yield was only 1·25%. What was the probable reason for such a very low yield?

6. Smith holds 340 Ordinary £1 shares and receives an income of £42·50 from them. What is the rate per cent. dividend?

7. Robinson sells £6400 Funding 6·5% Stock at 79. How much money does he receive? He reinvests the money in 5% Treasury Stock at 67½. Find by how much his annual income changes.

8. Williams sells 615 Marks and Spencer 25p Ordinary Shares at 139p. How many Distillers 50p Ordinary Shares can he purchase at 178p with the proceeds of the sale?

9. Briggs invested £354·20 in National Westminster Bank £1 Ordinary Shares at 253p. He sold them when they had risen to 330p. What percentage profit did he make on the sale? How much money did he make?

6. Expenses Involved in the Purchase and Sale of Securities

There are three items of expenditure involved in buying and selling stocks and shares, other than the actual cost of the securities. These incidental expenses usually total about 3·5% of the price of the shares, when £500 or more is involved. The expenses are as follows:

(1) Stamp Duty (i.e. cost of Transfer Stamp)
(2) Brokerage
(3) Contract Stamp.

Of these, the Contract Stamp charge is very small and can be neglected when working problems in this book, if desired. The worked examples do, however, take it into account.

For small transactions, the incidental expenses form an appreciably higher percentage of the price of the shares, for the minimum brokerage is £4.

6a. Stamp Duty

This is a duty payable to the Government when a transfer of stock (or shares) is made. This is approximately 2% of the consideration, i.e. of the cost of the stock (or shares).

TRANSFER STAMPS

Cost of Security	Duty Payable
Not exceeding £5	10p
Not exceeding £100	20p per £10
Not exceeding £300	40p per £20
*Exceeds £300	£1 per £50 (or fractional part thereof)
*EXAMPLE	
Transfer stamp on cost of £435	£400 = 8 × £50 Fractional part } £35 ∴ stamp £8+£1 = **£9**

NOTE

British Funds (e.g. Consols and War Loan) and most Public Corporation Stocks are free of Stamp Duty.

It will be observed that the Transfer Stamp normally costs slightly more than 2% of the amount spent on the security. This expense is borne by the buyer, not by the seller.

6b. Brokerage

A full table of the scale of minimum commissions payable to stockbrokers would be much too long for inclusion in a book of this type. Stockbrokers have an accepted scale of payments and a few examples are given below. The full list is easily obtained, if required, from any stockbroking firm or from a bank.

EXAMPLES OF SCALE OF MINIMUM COMMISSIONS

brought into effect recently.

Government Securities having no final redemption date within ten years.

0·625% on first £2000 consideration
0·250% on the next £12 000 consideration
0·125% thereafter, up to £986 000.

There is a ceiling on the brokerage charged for the transfer of securities of this class.

Shares or Units of Stock

1·5% on the first £7000 consideration
0·625% on the nest £93 000 consideration.

It is unlikely that the average reader will be interested beyond this level, for anyone with larger sums to invest will already have expert advice on investment!

The small investor is, however, again warned of one important point: there is a minimum brokerage of £4 on government securities, and £7 on other classes of stock or shares. Thus brokerage on a purchase of less than £467 ordinary shares would be higher than 1·5% (quoted above), for 1·5% of £467 = 7·001%.

Brokerage is paid by the buyer to his own broker and also by the seller to *his* broker.

6c. Contract Stamp

This is a stamp which must be affixed (by law) to a Contract Note. The procedure will be explained in Section 7 below.

Where the value of the shares or stock marketed costs:

From £100 to £500   10p ⎫
   £500 to £1500   30p ⎬ COST OF CONTRACT STAMP
   For amounts in
   excess of £1500   60p ⎭

7. Documents Involved in the Purchase of Securities

If we refer back to Section 2 above we see that Jones's broker has sold his shares at 273p. The broker will now immediately send Jones a *Contract Note* made out as follows:

CONTRACT NOTE 1

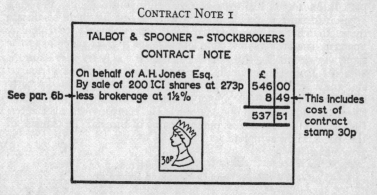

TALBOT & SPOONER – STOCKBROKERS
CONTRACT NOTE

On behalf of A. H. Jones Esq.
By sale of 200 ICI shares at 273p   £ 546 00
See par. 6b → less brokerage at 1½%   8 49 ← This includes
    537 51   cost of
    contract
    stamp 30p

30p

This is a guarantee to the seller that Talbot & Spooner will send the sum of £537·51 in due course when the transfer is completed.

In the meantime the shares have passed through the hands of the jobber to broker B (of the firm of Hart & Field, say), who buys them on behalf of Brown at a price of, perhaps, 275½p. Hart & Field send a contract note to Brown made out as shown in Contract Note 2.

CONTRACT NOTE 2[31]

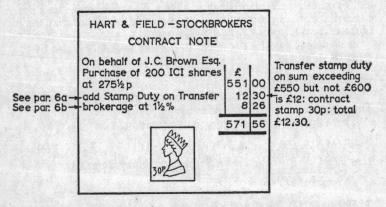

HART & FIELD – STOCKBROKERS
CONTRACT NOTE

On behalf of J.C. Brown Esq. Purchase of 200 ICI shares at 275½p	£	
	551	00
See par. 6a → add Stamp Duty on Transfer	12	30
See par. 6b → brokerage at 1½%	8	26
	571	56

Transfer stamp duty on sum exceeding £550 but not £600 is £12: contract stamp 30p: total £12.30.

It is necessary for Brown to retain this until the transfer is completed. As soon as Brown receives this contract note he sends £571·00 to Hart & Field. Jones will have sent his share certificate (proof that he had two hundred I.C.I. Shares for sale) to Talbot & Spooner. The brokers now complete a *Transfer Deed* which has to be signed by Jones. To this transfer deed the costly *transfer* stamp is affixed (£12 in this example; the remaining 30p being the *contract* stamp). When completed, it is sent with the share certificate formerly held by Jones to the company's registered offices (the Imperial Chemical Industries, in our example). The company cancels the certificate and issues a new one in the name of Brown. The transfer of the shares is now complete.

---

[31] Brokerage is not chargeable on stamp duty.

The whole of the business is illustrated diagramatically:

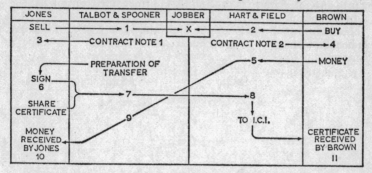

We shall now complete the chapter with some examples involving brokerage etc., and explanation of a few terms used.

8. Miscellaneous Examples

Ex. 5. Find the net amount realised when 250 Barclays £1 shares are sold at 364p.

	£
250 shares at 364p	910·00
Less brokerage (1·5%) £11·37 ⎫	13·95
+contract stamp 30p ⎭	
	896·05

Ex. 6. What is the total cost of 1000 Spear & Jackson Ordinary Shares at 128p?

	£
1000 at 128p	1280·00
Add brokerage at 1·5%	19·20
Contract stamp	0·30
Transfer stamp	26·00
	1325·50

Ex. 7. Smithers sells 500 Granada 'A' shares at 96p. He reinvests the money in Distillers 50p shares at 180p. How many does he get?

	£
500 at 96p	480·00
Less brokerage at 1·5% ⎫	7·30
+contract stamp 10p ⎭	
	472·70

We can approach the new purchase indirectly as follows. The stamp duty on £472·70 is £10 and the contract stamp is 10p, leaving £462·60.

The cost of each share is £1·80 plus brokerage of 1½% of this, i.e. £1·827. So we have

$$\text{Number of shares} = \frac{462·60}{1·827}$$

$$\fallingdotseq 253$$

$$
\begin{array}{r}
253 \\
1·872\overline{)462·60} \\
365·4 \\
\hline
97·20 \\
91·35 \\
\hline
5·850 \\
5·481 \\
\hline
·369 \\
\end{array}
$$

There would be a very small cash adjustment necessary.

In practice, from a set of tables, the approximate number of shares which could be purchased would be found, say 320 or 325 shares. The cash adjustment necessary would be made after the purchase of these shares. It would only be a few pounds one way or the other.

EXERCISE 2

1. Find the net amount realized on the sale of 800 Boots Pure Drug 25p shares at 203p.

2. Calculate the total cost of 600 Mars Transport 7% £1 Preference Shares at 88p. Allowing for tax at 34p in the £, find the net annual income from the shares.

3. Brown sells 400 Great Universal Stores 25p shares at 292p. Using the proceeds of the sale, he buys 350 Thorn Electrical 25p shares at 324p. How much cash does he owe?

4. Thompson sells 460 Guinness 25p Ordinary Shares at 136p and 600 Fisons £1 Ordinary Shares at 250p. Find how much he receives altogether. (Each transaction must be carried out separately.)

5. Jones invested £295 in J Lyons 'A' Ordinary £1 shares at 198p. Find the number of shares bought. If the net dividend (after deduction of tax at 34%) is 7·69p, find the net income received (before tax deduction) and the net yield per cent.

9. Some Types of Stocks and Shares

The reader will have noticed such names as Preference Shares and Ordinary Shares in this chapter. A short explanation will now be given.

Most of the stocks issued by the Government and by Local Authorities are fixed-interest holdings; that is, they pay a definite amount of interest each year. These stocks are quoted in multiples of £100, but they may be bought in units of £1, or even sometimes in units of 1p. For example, a certificate for £2174·68 of 3·5% War Loan would not be unusual.

Shares in companies are more varied. We shall divide them into two classes: I. Fixed-Interest Holdings; II. Equities.

There are hybrids of I and II as well.

I. Fixed-Interest Holdings

(a) *Debentures*. These are a mortgage holding secured on the assets of a company.

(b) *Preference Shares*. These are, in general, fixed-interest securities, but their dividend is paid out of profits before the ordinary shareholders are entitled to consideration. There are two *main* types of Preference Share: *Cumulative*—meaning that if the dividend is missed during a period of poor trading it must be made up later; *Non-cumulative*—meaning that if payment is not made in any year, there is no obligation to make it good later. Most preference shares are cumulative.

II. Equities

(a) *Ordinary Shares*. Dividends on these are paid out of the profits, after all expenses have been met and debenture and preference shareholders satisfied. They can be paid any dividend the directors think suitable, and the dividend frequently varies from year to year, or even more often.

Of other kinds of stock and share, and subdivisions of the above (e.g. Participating Preference Shares, Preferred Ordinary Stock, Prior Lien Debentures) there is not the space to deal.

10. Income Tax

The dividends and interests received from holdings of stocks and shares rank as unearned income, and no earned-income relief is allowable. The tax is deducted at the standard rate on almost all securities, before a cheque for dividend due is sent to a shareholder. Should the shareholder not earn enough to pay tax at the standard rate, he is entitled to claim back the difference between his rate of tax and the standard rate from the taxation authorities when he fills in his annual Income-Tax return.

Ex. 8. Roberts buys 2400 British Petroleum 8% Preference £1 Shares. If tax is deducted at 34p in the £, find the net annual income.

The gross income is £2400 $\times \frac{8}{100}$.

Now out of every 100p, tax of 34p is paid. The net amount received is therefore 66p out of every 100p.

∴ Net income is £2400 $\times \frac{8}{100} \times \frac{66}{100}$ = **£126·72.**

EXERCISE 3

1. Find the net income from £23 400 Gas 3% Stock (1990–95), if tax is deducted at 34p in the £.

2. British Oxygen 25p Ordinary shares pay 11% dividend. Find the income from 600 of these shares, if tax is deducted at 40p in the £.

3. Jenkins holds the following securities:

   250 National Westminster Bank £1 shares paying 17·5%
   360 Granada 'A' 25p shares paying 11·75%
   £1240 Funding 6% Stock (1993).

   If tax is deducted at 34p in the £, find the net annual income.

11. Authorised Unit Trusts

No chapter on investment would be complete without a brief mention of unit trusts, which in recent years have gained increasing popularity with the small investor.

The basic idea of the unit trust is to enable the small investor to purchase an interest in a large spread of companies, thereby substantially reducing his risk in any one business, which it would be beyond his means to do in any other way. Each trust is run by a professional management, whose job it is to see that, by careful purchase, sale and new purchase of various stocks and shares, the unit trust as far as possible enhances in value.

The small investor has nothing to do other than to write to the managers of a particular trust, asking to buy or to sell so many units. If he is buying, he encloses his cheque with his order. The expenses borne by the managers in these transactions are secured by a margin lying between their *offer* price and their *purchase* price. At the time of writing, for example, the Scottish Save and Prosper Group quote Scotbits, one of the best known unit trusts, at 39½p–42p. This means that they would *charge* an investor 42p for one unit or would *buy* from an investor at 39½p a unit. The margin between the prices (i.e. 2½p) would cover their expenses. There is also a small deduction, from the annual dividend, for running expenses and for the management's profit.

The dividend received from a unit trust tends to be low—sometimes very low. This would appear to be because for many people capital appreciation, rather than immediate income, is sought. Thus, managements often tend to buy investments with good prospects rather than good current returns.

Some examples of unit trusts should make the above points clear:

	Gross Yield %
National Westminster	
Extra Income 66·2p–71·1p	7·63
The Scottish Save and Prosper Group	
Scotbits 39½p–42p	3·80
Allied Hambro Group	
Pacific Fund 38·2p–40·9p	2·44

These are only three examples out of many. Each management group runs several different classes of unit trust. The investor does not necessarily seek the unit trust with the highest yield; he would probably be unwise to do so. He must decide in what class (e.g. capital increase, financial trust, high income, mining and metals, property, banking and insurance) he wishes to participate. Some investors choose more than one unit trust.

Ex. 9. Thomas has £280 to invest in buying a whole number of units in Abbey Unit Trust Mgrs. Ltd. Income, currently quoted at 39p–42p. How many units does he get?

The rate of yield is 5·53%. What is his gross income?

Thomas will have to pay 42p a share. He therefore acquires

$$\frac{28\ 000}{42} = 666\tfrac{2}{3} \text{ shares, i.e. } \mathbf{666\ shares} \text{ are bought.}$$

The income received is £280 × $\dfrac{5·53}{100}$ ⌒ **£8·96.**

Note that the income is quoted as *gross percentage yield on outlay* on the day concerned. It is not given in the press as a percentage dividend.

Exercise 4

1. Norwich Union Insurance Group Trust Fund is quoted at 345·3p–363·5p. If Smith sells 740 units, how much money will he get?

2. Hill Samuel Unit Trust Mgrs. International Trust is quoted at 38p–40·7p, the yield being 3·17%. Richards invests £162 to buy a whole number of units. What will be his gross annual income?

   If later, he is obliged to sell his units at the same ruling price range shown above, how much will he receive?

3. Clark buys 540 units of Lloyds Bank First Income Trust, quoted at 50·5p–54·3p and yielding 4·34%. He later sells them when the price has risen to 62·5p–66·3p. How much profit does he make on the transaction?

During the time when Clark holds the units, what is his net annual income, after deducting tax at 42p in the £?

4. Stevens sells 10 000 Barclay Unicorn Ltd. Recovery Trust, quoted at 42p–45·4p. He reinvests the proceeds in 5·5% Treasury Stock (2008–12) at 49. Find how much stock, to the nearest £10, he buys. What will be his new gross annual income? (Do not forget to allow for *brokerage* on the purchase of Treasury Stock, although it is free of stamp duty.)

CHAPTER SEVENTEEN

VALUE ADDED TAX (V.A.T.)

1. Definition and Description of its Application

The definition of V.A.T. is that it is a tax levied on the increase in the value of a product, arising from manufacturing and marketing processes. As the tax closely affects the purchase or sale of so many kinds of goods, virtually every reader of this book must be aware of its existence, but there may be some puzzlement as to why earlier treatment has not been given to this important topic. The reason is that had the tax been included where applicable, a large number of the exercises involving money would each have needed appreciable extension, which would have been irrelevant and confusing from the point of view of mastering new ideas.

It therefore seemed better to concentrate on the arithmetical processes directly concerned and finally to present a short commentary on Value Added Tax. For the ordinary customer the tax presents no difficulties, but in reality the general application of the levy is complex and frequently presents serious problems to manufacturers and traders. One hears of shopkeepers having to spend hours each evening in order to keep their Customs and Excise accounts in order, as a result of the introduction of V.A.T. a few years ago. It would seem, also, that many small businesses have had to close because of the difficulties involved.

At the present time (1978), some categories of commodity are exempt from V.A.T., the most important being a large group of foodstuffs, but excluding, for example, concentrated soft fruit drinks, chocolate and ice-cream. Most fabricated goods are rated currently at 8%, but certain 'luxury' goods, including many kinds of electrical appliances, together with items made, say, of gold and silver, are charged at $12\frac{1}{2}\%$.

Shopkeepers usually state a price inclusive of V.A.T., but elsewhere—in perhaps a hotel or restaurant—the tax is commonly added separately to one's bill (and may even become entangled with service charge!).

2. Simple Applications of V.A.T.

The general application of the tax is best illustrated by a typical process in the manufacture and sale of an everyday article.

Consider what happens when a wooden chair proceeds from a supplier of timber to a private customer. The chain of transactions, excluding minor items such as nails, screws, glue and polish, may take a form something like this:

Supplier (S)→Manufacturer (M)→Wholesaler (W)→Retailer (R)→ Customer (C)

There are four steps involved:

(1) M pays S his price, C_1, plus 8% V.A.T., and S pays this amount in excise,

$$\text{i.e. S pays } \frac{8}{100} C_1$$

(S may have purchased tree trunks and reduced these to useful and workable timbers).

(2) W pays M his price, C_2, which will be higher than C_1, plus 8% V.A.T. on C_2; M then pays $\frac{8}{100} C_2$ in excise, and claims back $\frac{8}{100} C_1$, *which he has already paid S*, from the excise department.

$$\text{Hence M pays } \frac{8}{100} (C_2 - C_1)$$

(3) R pays W his price, C_3, which will be higher than C_2, plus 8% V.A.T. on C_3; W then pays $\frac{8}{100} C_3$ in excise, and claims back $\frac{8}{100} C_2$, which he has already paid W, from the excise department.

$$\text{Hence W pays } \frac{8}{100} (C_3 - C_2)$$

(4) C pays R his price, C_4, which will be higher than C_3, plus 8% V.A.T. on C_4; R then pays $\frac{8}{100} C_4$ in excise, and claims back $\frac{8}{100} C_3$, as in stages (2) and (3) above.

$$\text{Hence R pays } \frac{8}{100} (C_4 - C_3)$$

The total tax paid on the chair is therefore

$$\frac{8}{100} C_1 + \frac{8}{100} (C_2 - C_1) + \frac{8}{100} (3_3 - C_2) + \frac{8}{100} (C_4 - C_3)$$

$$= \frac{8}{100}(C_1+C_2-C_1+C_3-C_2+C_4-C_3) = \frac{8}{100}C_4,$$

as all of the other terms cancel out.

Therefore, the total tax paid is 8% of the final (retail) price, C_4.

Poor C, the customer, does not get any tax rebate from the excise department! In effect he has paid the whole lot.

Ex. 1. The jewellery firm of T. Jones wishes to sell a gold bangle for £36, plus V.A.T. which is rated at $12\frac{1}{2}\%$. What is the selling price?

Basic price	£36·00
$12\frac{1}{2}\%$ V.A.T.	4·50
Total price	**£40·50**

The purchaser will have to pay **£40·50**.

Ex. 2. Mrs Brown buys 2 lb of strawberries at 35p per lb, 3 bottles of orange squash at 32p each, 5 litres of ice-cream for £1·85 and a 500 g packet of cornflakes for 38p. If V.A.T. is added, and strawberries and cornflakes are rated at 0% V.A.T., whereas the orange squash and ice-cream are at 8% V.A.T., find Mrs Brown's total bill.[32] (This example illustrates the shambles to which our weights and measures have been reduced. See p. 62.)

	Net Cost (£)	V.A.T. (£)
2 lb strawberries at 35p/lb	0·70	0·00
3 bottles of squash at 32p each	0·96	0·08
Ice-cream	1·85	0·15
Cornflakes	0·38	0·00
	3·89 +	0·23 = **4·12**

Mrs Brown pays **£4·12** in all, of which **23p** is Value Added Tax.

Ex. 3. George Smith, a retailer of decorative materials, pays £75, plus V.A.T. at 8%, for 60 tins of paint. He wishes to make a profit of 50p a tin. At what price does he sell it? (We shall use this example to show that the application of V.A.T. does not affect Smith's profit, although it certainly affects the price that the private customer must pay).

Consider firstly the sale of the tin of paint *without any V.A.T.*

[32] In some large establishments and warehouses, *V.A.T. is added* to the market price.

Cost price of one tin $£\dfrac{75}{60} = £1\cdot25$

Profit $0\cdot50$

Selling Price $£1\cdot75$

Now consider the sale of the tins of paint, when V.A.T. is taken into account.

Smith paid his wholesaler £1·25, plus 8% V.A.T., for each tin of paint.

	$£$
Basic cost	$1\cdot25$
8% V.A.T.	$0\cdot10$ (this is called his *input tax*)
Total cost	$1\cdot35$

Smith wishes to get 50p actual profit, and hence he charges

	$£$
Basic cost	$1\cdot25$
Profit	$0\cdot50$
	$1\cdot75$
8% V.A.T.	$0\cdot14$ (*output tax*)
Total price	$1\cdot89$

Now the tax Smith has to pay is £0·14 and the tax he reclaims is £0·10 (i.e. he owes Output Tax—Input Tax). He therefore pays £0·14—£0·10 = £0·04, which is in accordance with the general result shown earlier; it corresponds to $\dfrac{8}{100}\,(C_4-C_3)$, where $C_4 = £1\cdot75$ and $C_3 = £1\cdot25$.

Hence Smith's profit is $£(1\cdot89-1\cdot35-0\cdot04) = £0\cdot50$, i.e. he gets 50p profit, after payment of the difference of 4p between output and input tax. This is exactly what he would have got if V.A.T. did not exist. The customer does not fare so well—he pays £1·89 instead of £1·75.

Ex. 4. Thompson asks his stockbrokers to sell £4000 of $9\frac{1}{2}$% Treasury Stock (1999). The best price the brokers can get is $80\frac{3}{8}$. What is the net amount Thompson will receive, after allowing for commission, V.A.T. and contract stamp? (N.B. There is no Transfer Stamp Duty as this is a *sale* of stock, not a purchase: see p. 176 *et seq.*)

The point of this question is the inclusion of V.A.T. *which only applies to the stockbroker's commission.* The procedure follows the lines of the example in Section 7 on p. 177, with the addition of the one

extra charge. Although laid out in full for clarity, it would normally
be briefly presented on a Contract Note (as on p. 178, but somewhat
extended).

		£	£
By sale of £4000 stock at 80⅜			3215·00
less (1) Commission {0·625% on £2000 cash		12·50	
0·25% on £1215 cash		3·04	
		15·54	
(2) V.A.T. 8% of brokerage (£15·54)		1·24	
(3) Contract Stamp (see p. 177)		0·60	
Total deductions		17·38	17·38
Net amount payable to Thompson			**£3197·62**

EXERCISE 1

1. A shopkeeper sells a ladder at a net price of £24·50. If V.A.T. is added
at 8%, what is the price a customer must pay?

2. A housewife buys some goods at a warehouse, where V.A.T. is added
to the cost price where relevant. The items bought are ½ lb bacon at
96p per lb, 1 lb chocolates for £1·24, 4 lb sugar at 28p for a 2 lb packet,
and 2 bottles of lemon squash at 34p each. The bacon and the sugar
are rated at 0% V.A.T., and the other items are at 8% V.A.T. Find
the total cost of the purchases and also the percentage of this which
is Value Added Tax.

3. Robinson buys two dozen identical radios for £876 from his wholesaler, V.A.T. being added at 12½%. If Robinson wishes to make £12
profit on each one, find the retail price which a customer must pay for
a radio. Also find the difference between Robinson's output tax and
input tax.

4. Williams buys £6000 Exchequer 10½% Stock (1997) at 87. Find the
total account rendered to him by his stockbrokers, if 8% V.A.T. is
included in the brokers' commission. (N.B. As the stock is a British
Fund, there is no Transfer Stamp Duty, even although this is a *purchase*: see p. 176, note).

5. A customer pays £73·71, inclusive of 8% V.A.T., for a sideboard.
What would have been the cost had it been free of Value Added Tax?
(p. 79, Ex. 5, is based on similar reasoning.)

This chapter is merely a simple introduction to straightforward
applications of V.A.T., which may perhaps help the layman to understand a little of what is involved. Many problems arising during the
operation of the tax have been omitted, such as the numerous forms
and returns which need detailed completion. No mention has been
made of antiques and original works of art, for which a 'special

scheme' is to some extent in operation, nor has there been reference to the regulations concerning the duty on imports and exports, arising from V.A.T.

The reader who wishes to gain more accurate and detailed information is referred to the long series of booklets issued by the Customs and Excise Department, published by H.M. Stationery Office, and dealing exclusively with Value Added Tax.

CHAPTER EIGHTEEN

COMPUTING MACHINES AND THE BINARY SCALE

1. Calculating Machines

Calculating machines have been in existence for a considerable number of years. Simple desk machines were made as early as the eighteen-eighties, but it is only recently that high-speed electronic machines of great efficiency have been devised. To a layman these are still a mystery, and they are indeed of very complex construction. The arithmetic of this chapter will be devoted to a particularly important machine, the *digital computer*, in which the basic operation is counting. Such a machine can perform the operations of addition, subtraction, multiplication and division and many more complex processes. Early digital computers were mechanical, but the latest models are almost entirely electrical. Teleprinters are often incorporated in these machines.

In addition to the digital computer there is another type of machine called the *analogue machine*, or continuous operator. This is a mathematical instrument in which there is a *continuous variation* of some physical quantity such as an electrical resistance. This is in sharp contrast to the basis of the digital computer, wherein the steps are of *one unit*, not of continuity. (A very simple example of an analogue machine is the slide rule, in which two lengths, representing two numbers to be multiplied together, are placed end to end and added. As any two numbers will give any two lengths, there is a continuous variation possible.) Analogue machines are often used for solving differential equations[33] such machines sometimes being given the name of 'differential analysers'. Analogue machines are taken to pieces rather like a fantastic Meccano set, and reassembled differently for each different kind of problem. Digital computers, on the other hand, once built need not be modified. A particular programme fed into a digital machine may not require the utilisation of more than a limited part of the total access available. As analogue machines involve mathematics of a complex nature unsuited to a book on simple arithmetic we shall not consider them in further detail.

A digital computer consists of a vast collection of circuits through which electric currents can flow. This flow, originally regulated by thermionic valves, is now controlled by transistors. Difficulties in

[33] See *Calculus*, P. Abbott, Teach Yourself Book.

utilising transistors are considerable, as their characteristics vary widely when temperature changes are small. If the impulse received is sufficiently large a current passes through a particular circuit: if not, no current flows in this section of the network. This implies the choice of two, and only two, possibilities—Yes or No.

Now this is the same as counting in the *scale of two* (the binary scale), for the only basic numbers which are permitted are o (no current) and 1(current flowing).

It is therefore clear that, for such a machine to work, it is necessary that a problem to be solved shall be converted into the binary scale, and the answer at the end shall be converted back. Before we actually carry out some calculations in the binary scale, it is worth spending a few moments considering whether the popular name 'electronic brains' applied to such machines has in fact any justification.

There is no structural similarity between the nervous system and computing machines, but there is a remarkable resemblance between a direct nervous impulse and the action of a transistor or a relay. In each case there is or is not a result, and the magnitude of this result does not vary. Computing machines calculate as we do, but they are vastly more efficient, for they are very much quicker and far more accurate. The elements of the brain are to some extent independent of one another, so in fact are the circuits of an electronic machine. The machine can store information fed into it and reissue that information as required—this is a kind of memory. There is little likelihood of devising machines of such complexity as the brain, which has some 10^{10} cells, whereas the biggest digital computers only have, perhaps, 10^5 separate circuits.

2. Binary Scale

When we use the scale of ten (the *denary scale*) we count from o to 9, but when we reach ten, we put a *one* in the tens column thus, 10, and start again 10, 11 . . . 99. When we reach one hundred, which is ten times ten, we put a *one* in the hundreds column thus, 100, and carry on 100, 101, 102. . . . As already explained in Chapter 2, the number 5278 is really

$$5 \times 1000 + 2 \times 100 + 7 \times 10 + 8,$$
$$\text{i.e. } 5 \times 10^3 + 2 \times 10^2 + 7 \times 10 + 8.$$

Similarly $2030 = 2 \times 10^3 + 0 \times 10^2 + 3 \times 10 + 0.$

Now the *binary scale* is obtained in exactly the same way. A number is expressed as powers of 2. Consider the number 43 of the ordinary decimal (denary) scale.

Now $43 = 1 \times 2^5 + 0 \times 2^4 + 1 \times 2^3 + 0 \times 2^2 + 1 \times 2 + 1.$

Therefore, in the binary scale 43 is 101011, for every time we reach

2 we proceed to the next column (e.g. $3 = 1 \times 2 + 1$, so in the binary scale it becomes 11).

We do not read a number 101011 in the binary scale as 'one hundred and one thousand and eleven'. As no separate nomenclature has been given to this scale of numbers, we say in words that the number 101011 is 'one nought one nought one one'.

How then do we calculate these strange new numbers? It is really very simple. We merely divide by 2 repeatedly, thus:

2)43
2)21+1
2)10+1
2) 5+0
2) 2+1
  1+0

The answer is read round the edge as **101011**.
Notice the remarkable resemblance to the method of reading off the answer in 15th-century arithmetic in Chapter 1.

The number of digits in a number expressed in the binary scale is about three times as many as that in the denary scale. This does not imply any difficulty of manipulation. In fact, nothing could be easier than the addition and multiplication tables. They are completely expressed as follows:

Addition Table

$$0+0 = 0 \quad \text{(nought)}$$
$$1+0 = 1 \quad \text{(one)}$$
$$1+1 = 10 \quad \text{(one nought)}$$

Multiplication Table

$$0 \times 0 = 0 \quad \text{(nought)}$$
$$1 \times 0 = 0 \quad \text{(nought)}$$
$$1 \times 1 = 1 \quad \text{(one)}$$

The immediate reaction might well be that there would be a great welcome in junior schools for such a system which would save so much of the grind of simple arithmetic. Perusal of the following examples may lead to a modification of that first impression.

We shall now apply our new system to the 'four rules' of arithmetic.

Ex. 1. Add together 17, 6 and 23 in the binary scale.

2)17   2)6   2)23
2) 8+1   2)3+0   2)11+1
2) 4+0   1+1   2) 5+1
2) 2+0   2) 2+1
  1+0   1+0

17	10001
6	110
23	10111
46 (decimal)	**101110** (binary)

Notes (a) In the binary scale $1+1 = 0$ and carry 1.
(b) $\qquad\qquad\quad 1+1+1 = 1$ and carry 1.

To convert back to the denary scale, when necessary, is merely to reverse the reasoning:

$$101110 = 1 \times 2^5 + 0 \times 2^4 + 1 \times 2^3 + 1 \times 2^2 + 1 \times 2 + 0$$
$$= 32 \quad + \quad 0 \quad + \quad 8 \quad + \quad 4 \quad + \quad 2 \quad + 0$$
$$= \mathbf{46.}$$

Ex. 2. Multiply 23 by 48 in the binary scale.

$48 \times$	$110000 \times$
23	10111
144	110000
96	110000
1104	110000
	000000
	110000
	10001010000

Notice that, had we reversed the order of multiplication, much labour could have been saved by filling in the four zeros first:

$$10111 \times$$
$$110000$$
$$\overline{}$$
$$101110000$$
$$10111$$
$$\mathbf{10001010000}$$

Converting back

$$10001010000 = 2^{10} + 2^6 + 2^4$$
$$= 1024 + 64 + 16$$
$$= \mathbf{1104.}$$

Ex. 3. Multiply 243 by 47 in the binary scale.

This example is given to illustrate the magnitude of the task when large numbers are manipulated. It also points out the method of handling larger carrying figures.

$$
\begin{array}{r}
243\times \\
47 \\
\hline
1701 \\
972 \\
\hline
\mathbf{11421}
\end{array}
\qquad
\begin{array}{r}
1111 0011\times \\
101111 \\
\hline
1111 0011 \\
1111 0011 \\
1111 0011 \\
1111 0011 \\
1111 0011 0 \\
\hline
\mathbf{10110010011101} \\
12333321111
\end{array}
$$

Checking back the result:
$$
\begin{array}{rcr}
2^{13} &=& 8192 \\
2^{11} &=& 2048 \\
2^{10} &=& 1024 \\
2^{7} &=& 128 \\
2^{4} &=& 16 \\
2^{3} &=& 8 \\
2^{2} &=& 4 \\
1 &=& 1 \\
\hline
& & \mathbf{11421}
\end{array}
$$

It is now clear why the system, fundamentally so simple, is only really suitable for accurate high-speed calculating machines. The multiplication of two five-figure numbers in the denary scale is likely to become fifteen rows of fifteen-figure numbers in the binary scale, involving an answer of some thirty digits.

It might be noticed in passing that the carrying figures placed underneath the appropriate columns in the above example are in the denary notation. We can use this (although it will not be written hereafter in this chapter) if we are careful to divide by 2, thus:

$1+1+1+1+1 = 1$ and carry 2 (i.e. two *ones*) to the next column.
We end, in the last column, with
$\qquad 1+1 = 0$, and carry 1
$\qquad\quad\; = 10$ (one-nought).

Ex. 4. Divide 2077 by 31 in the binary scale.

$$
\begin{array}{r}
\mathbf{1000011} \\
11111\overline{)10000001 1101} \\
11111 \\
\hline
101110 \\
11111 \\
\hline
11111 \\
11111 \\
\hline
\cdots\cdots
\end{array}
$$

Converting the answer back to the denary scale

$$1000011 \equiv 2^6 + 2 + 1 = 67.$$

EXERCISE 1

1. Convert the following denary numbers to binary numbers:

　　(*a*) 14　　(*b*) 34　　(*c*) 205　　(*d*) 1101　　(*e*) 4758.

2. Convert the following binary numbers to denary numbers:

　　(*a*) 110　　(*b*) 11　　(*c*) 10101　　(*d*) 100110　　(*e*) 11100101011.

3. Add together the following numbers after first converting them to the binary scale. Give the answers in (i) binary scale, (ii) denary scale.

　　(*a*) 6+19+8　　　　　　　(*b*) 21+43+7
　　(*c*) 103+68+274　　　　(*d*) 3024+953+67+118.

4. Subtract 214 from 316 in the binary scale, giving the answer in binary form.

5. Add the following binary numbers, and afterwards convert the result to denary form: 1101, 11001, 100101, 101110.

6. Evaluate the binary number sum

$$1101 - 100111 + 111000.$$

7. Multiply the following numbers after converting them to binary form. Give the answers in (i) binary scale, (ii) denary scale.

　　(*a*) 23×9　　(*b*) 68×27　　(*c*) 104×35
　　(*d*) 214×6×7　　(*e*) 109×31×14.

8. Divide 2173 by 18 in the binary scale. Give the answer in the scale of two, and then convert it to the scale of ten.

9. Divide 11011101 by 10011, both numbers being in the binary scale. Give the result in (i) binary form, (ii) denary form.

The future of computing machines is a great one. Problems which involve prodigious quantities of arithmetic, involving possibly years of work by a man, can be done by machines of this kind in a few minutes.

During the past few years, in addition to the very costly full-scale computers there have appeared in ever increasing numbers the most useful pocket-sized electronic calculators. These were initially fairly expensive, say £150 for a good one. Now, however, £20 to £30 will purchase quite elaborate instruments, whilst small capacity ones are obtainable from about £5 upwards. The more complex calculators have virtually sounded the knell for slide rules and logarithms, in so far as most people are concerned, for these new machines are much faster in operation and, in general, are more accurate—if used correctly!

The fundamental difference between a computer and a calculator is that, whereas the former is programmed once only, for a *specific* series of operations which may be repeated indefinitely for different sets of arithmetical values, the latter needs to have a key depressed for *every single operation* as well as for each new group of numbers. Furthermore the *pocket* calculators (as distinct from some more expensive desk types) do not have a paper feed-out, but merely give results as groups of illuminated numbers. These limitations greatly reduce the size and cost of the machines.

As will have been seen earlier (p. 159), there is a certain minimal list of operations needed for a pocket calculator to be fairly generally useful.

Those who would like to gain further understanding of the arithmetic and logic of computers may wish to continue by reading *New Mathematics* (in the Teach Yourself Books Series), by the present author. The book follows on naturally from *Arithmetic (Decimalized and Metricated)*.[34]

[34] Teach Yourself Books—*New Mathematics*, Chapter 4, extends the ideas of the binary scale and introduces other scales.

ANSWERS to EXERCISES

ANSWERS

CHAPTER 2

EXERCISE 1

1. 43   2. 42   3. (a) 370 (b) 2415 (c) 13 539
4. £291, £288, £312, £288, £243, £342; total wages bill £1764

EXERCISE 2

1. (a) 16   (b) 45   2. (a) 130   (b) 49   3. 66
4. (a) + +   (b) − +   (c) + −   (d) − −

EXERCISE 3

1. (a) 85   (b) 161   (c) 1512   (d) 2354
2. (a) 408   (b) 2502   (c) 31 400   (d) 114 948
3. (a) 5072   (b) 11 856   (c) 38 617   (d) 944 672
4. 5796   5. 289 085

EXERCISE 4

1. (a) 353   (b) 246   (c) 8069, rem. 1
2. 954 m   3. 38   4. 2555   5. $48\frac{1}{13}$

EXERCISE 5

1. (a) 56   (b) 103   (c) 580
2. (a) 226   (b) 1238   (c) 7088   (d) 65 124
3. (a) 161   (b) 2067
4. (a) 388   (b) 2127
5. 1996   6. 3829
7. (a) 1953   (b) 782   (c) 24 273   (d) 92 117
8. (a) 9976   (b) 22 192   (c) 51 330   (d) 733 824
9. (a) 12   (b) 37   (c) 293   (d) 4178, rem. 5
10. (a) 49 rem. 41   (b) 38   (c) $36\frac{39}{88}$ (d) $151\frac{90}{133}$
11. (a) 77   (b) 23   (c) 38   (d) 17
12. (a) 27   (b) 5
13. 55 380   14. 45 years   15. 519 metric tons
16. 117 rows (including one incomplete row containing 55 cabbages)
17. 2304   18. 8784 (366 days in 1984)   19. 4784
20. 63 kilometres   21. 86 400
22. The error is $(1732-1723) \times 27 = 9 \times 27 = 243$
23. (a) 450   (b) 599   (c) 300   24. £794

CHAPTER 3

EXERCISE 1

1. (a) 2   (b) 2, 3, 6   (c) 2, 3, 5, 6   (d) 2, 3, 4, 5, 6   (e) 5
2. (a) Divisible by both   (b) Divisible by both
   (c) Divisible by 9 but not by 8

3. (a) $2 \times 3 \times 3$ (b) $2 \times 2 \times 3 \times 5$
   (c) $2 \times 2 \times 3 \times 7 \times 11$ (d) $3 \times 7 \times 11 \times 13$
4. (a) $2^4 \times 3$ (b) $3^2 \times 7^2$
   (c) $2^4 \times 3 \times 11^2$ (d) $2^3 \times 5 \times 7^2 \times 11$
5. (a) 43   (b) $7\overline{1}$ (c) $35\frac{8}{11}$ (d) $338\frac{22}{27}$

EXERCISE 2   *page 23*

1. (a) 21   (b) 8   (c) 22   (d) 36
2. (a) 24   (b) 24   (c) 108   (d) 4840
   (e) 60   (f) 252   (g) 560
3. 1320   4. £6
5. 12 m = 1200 cm; H.C.F. of 1200 and 175 is 25
   ∴ tiles are 25 cm by 25 cm; number required 336
6. 72   7. L.C.M. of 1, 2, 3 ... 10 = 2520   8. 16

CHAPTER 4

EXERCISE 1   *page 26*

1. 16 h   2. 48 min   3. 375 cm³   4. 84 s (= 1 min 24 s)
5. 6 min 24 s   6. 87½p   7. $\frac{1}{14}$ 8. $\frac{1}{6}$
9. $\frac{1}{40}$ 10. $\frac{5}{9}$ 11. $\frac{5}{9}$

EXERCISE 2   *page 29*

1. $1\frac{1}{4}$ 2. $\frac{1}{8}$ 3. $4\frac{1}{10}$ 4. $1\frac{1}{8}$
5. $1\frac{3}{4}$ 6. $6\frac{1}{12}$ 7. $1\frac{17}{20}$ 8. $2\frac{11}{42}$
9. $\frac{29}{60}$ 10. $\frac{24}{25}$ 11. $1\frac{31}{32}$ 12. $1\frac{1}{8}$

EXERCISE 3   *page 31*

1. $\frac{8}{21}$ 2. $\frac{5}{8}$ 3. $3\frac{2}{5}$ 4. $\frac{9}{16}$
5. $35\frac{3}{4}$ 6. 2   7. $\frac{1}{8}$ 8. $1\frac{1}{8}$
9. $2\frac{2}{3}$ 10. $1\frac{1}{2}$ 11. $\frac{1}{2}$ 12. $1\frac{23}{147}$
13. $\frac{9}{25}$ 14. 2

EXERCISE 4   *page 32*

1. $1\frac{4}{15}$ 2. 17   3. $\frac{42}{205}$ 4. $\frac{2}{85}$
5. $1\frac{11}{32}$ 6. $\frac{7}{17}$

EXERCISE 5   *page 33*

1. $\frac{3}{8}$ 2. £4500   3. £1125   4. $14\frac{4}{19}$
5. 17 days   6. 24 km   7. 56¼ min   8. £2·92½

CHAPTER 5

EXERCISE 1   *page 37*

1. (a) 2·3   (b) 6·01   (c) 37·57   (d) 0·91
   (e) 0·07   (f) 0·0004   (g) 85·076
2. (a) $6+\frac{8}{10}$ (b) $10+4+\frac{9}{10}+\frac{2}{100}$ (c) $200+7+\frac{4}{100}$
   (d) $\frac{6}{10}$ (e) $\frac{9}{100}+\frac{1}{1000}$ (f) $1000+\frac{1}{1000}$
3. (a) 31·262   (b) 330·614   (c) 2013·5   (d) 1470·8004
4. (a) 23·7   (b) 13·75   (c) 9·073   (d) 297·9155

EXERCISE 2   *page* 41

1. 417   2. 0·087   3. 22·010   4. 6·1
5. 0·0070   6. 0·0309   7. 9·84   8. 0·098

EXERCISE 3   *page* 42

1. (*a*) 0·4   (*b*) 63   (*c*) 3850   (*d*) 7
   (*e*) 0·15   (*f*) 7·08   (*g*) 1·6   (*h*) 492
   (*i*) 5·04   (*j*) 2023   (*k*) 222·64   (*l*) 0·0053
2. (*a*) 3·4   (*b*) 1·45   (*c*) 0·325   (*d*) 0·0008
   (*e*) 0·192   (*f*) 1·14   (*g*) 0·193   (*h*) 5·5
   (*i*) 5400   (*j*) 180   (*k*) 0·003 652   (*l*) 28·3
3. (*a*) 93·86   (*b*) 0·2856   (*c*) 27·8064   (*d*) 11·22
   (*e*) 2500   (*f*) 507   (*g*) 3599·8   (*h*) 17·60   (*i*) 14·23
4. (*a*) 0·89   (*b*) 0·073   (*c*) 1·39   (*d*) 22·0
   (*e*) 0·0084   (*f*) 22·470

EXERCISE 4   *page* 43

1. (*a*) 0·6   (*b*) 0·625   (*c*) 0·6364   (*d*) 0·2222
   (*e*) 0·3889   (*f*) 0·5294
2. (*a*) $\frac{3}{25}$ (*b*) $\frac{179}{250}$ (*c*) $\frac{17}{200}$ (*d*) $\frac{453}{5000}$ (*e*) $7\frac{13}{80}$
3. $\frac{427}{125} = 3·416$, $\frac{147}{43} = 3·419$, $\frac{171}{50} = 3·420$
4. (*a*) $\frac{1}{25}$ (=0·04)   (*b*) 54
5. (*a*) 6·4516   (*b*) 0·0004   (*c*) 0·001 331
6. (*a*) 0·18   (*b*) 10

CHAPTER 6

EXERCISE 1   *page* 47

1. (*a*) 2000 m   (*b*) 3750 m   (*c*) 41 500 m   (*d*) 400 m
   (*e*) 0·07 m   (*f*) 0·38 m   (*g*) 0·0576 m   (*h*) 6·5 m
   (*i*) 0·006 m   (*j*) 0·082 m   (*k*) 0·076 m   (*l*) 0·457 m
3. (*a*) 48, 64, 80, 112 km/h   (*b*) 13·3, 17·8, 22·2, 31·1 m/s
4. £65·62   5. 86·9 m   6. 9 s

EXERCISE 2   *page* 49

1. (*a*) 3 000 000 m²   (*b*) 1 450 000 m² (*c*) 307 000 m²   (*d*) 120 000 m²
   (*e*) 0·05 m²   (*f*) 0·0512 m²   (*g*) 0·000 362 m² (*h*) 0·75 m²
   (*i*) 0·000 008 m²   (*j*) 4700 m²   (*k*) 0·082 46 m²
3. 0·512 ha   4. 4046 m²   5. 42 m

EXERCISE 3   *page* 52

1. (*a*) 4800 cm³   (*b*) 254 800 cm³ (or 254·8 dm³)
2. 378 l   3. 0·454 kg   5. 1·76 pint   6. 1990 cm³
7. 3 000 000 m³   8. 1360 bricks   10. 1·76 kg/cm²   11. 6·67 km

CHAPTER 7

EXERCISE I *page* 60

1. 62p   2. £4·76   3. £254·02   4. £1·95
5. £15·07   6. 97½p   7. 8½p   8. £3·88½
9. 70½p (nearest ½p)   10. £1·36

EXERCISE 2 *page* 61

1. £57·80   2. £15·12   3. £15·91   4. £84·49½
5. £52   6. £97   7. 64p   8. 34p
9. £6·24   10. 51p   11. £1711·85   12. £239·23
13. £1412·20   14. £52·50   15. 15 shears; £1·75
16. 3/4   17. 53 torches; 30½p   18. £3928
19. £2·67½   20. £987·36

EXERCISE 3 *page* 63

1. 3 mi 1660 yd   2. (a) 16 428 lb   (b) 7450 kg   3. 27·90 ton
4. (a) 75 ac   (b) 30·36 ha   5. 1·23 pt; 1·32 pt; 1·76 pt

EXERCISE 4 *page* 64

1. £7·96   2. £7·80   3. £10·73   4. £53·43
5. £131·35   6. £166·24   7. £50·22½

EXERCISE 5 *page* 65

1. £3·15   2. 58p   3. 1·38..; 1·31..; the second

CHAPTER 8

EXERCISE I *page* 67

1. 3:50   2. 1·62:1 (3 s.f.)   3. 25:28   4. 3:4
5. 5:3   6. £13   7. 1·35:1   8. 4:3
9. 0·4375   10. 60   11. £3216; £2412; £1608   12. 7:6

EXERCISE 2 *page* 71

1. £97·20   2. 16 min   3. £1920   4. 37½ days
5. £676·20   6. 46 men

EXERCISE 3 *page* 72

1. £20, £16·66½, £10, £3·33½ (nearest ½p)
2. £100 each man; £75 each woman
3. £30, £45, £54   4. The first   5. 17½ min   6. 17:9; £76·50

EXERCISE 4 *page* 74

1. £632   2. (a) 5:14   (b) 0·143:1   (c) 261p
3. 36 min   4. 36·3 approx   5. 67·6 km/h   6. 30 runs
7. 23 405; 52 662

CHAPTER 9

EXERCISE 1   *page* 77

1. (a) 20%   (b) 25%   (c) $66\frac{2}{3}\%$ (d) $63\frac{7}{11}\%$
   (e) 112·5%   (f) 0·429% (approx.)
2. (a) $\frac{3}{20}$ (b) $\frac{7}{25}$ (c) $\frac{11}{40}$ (d) $1\frac{3}{5}$
   (e) $\frac{1}{3}$ (f) $\frac{1}{250}$
3. (a) 0·35   (b) 0·3333   (c) 1·075   (d) 0·0375
   (e) 0·0164   (f) 0·0343
4. (a) 31·11%   (b) $12\frac{1}{2}\%$ (c) 8·46%   (d) 7%
5. (a) £9·00   (b) £1·91   (c) 520 cm³   (d) £7·92

EXERCISE 2   *page* 81

1. (a) £4·80   (b) £4·35   (c) 55p
2. (a) $12\frac{1}{2}\%$ profit   (b) 25% loss   (c) $31\frac{1}{4}\%$ profit
3. (a) £2·50   (b) £4   (c) 99p
4. (a) $19\frac{1}{2}$p   (b) £1·90   (c) $£46·70\frac{1}{2}$
5. 0·92%   6. £216   7. 3·88%   8. 0·23%; $£48·87\frac{1}{2}$

EXERCISE 3   *page* 84

1. £1473·90   2. $7\frac{3}{4}\%$ 3. 8%   4. £14·96
5. 9·375%   6. £3·6; 37·1%   7. 10·53%
8. £3512·88; 12·74%

EXERCISE 4   *page* 85

1. 5:7   2. £3·22   3. 20·7%   4. $33\frac{1}{3}\%$
5. 4:3   6. 31   7. 22

CHAPTER 10

EXERCISE 1   *page* 91

1. 16 m; 2·2 s   2. 11.20 a.m.; 225 km
3. £8400; advertising costs eventually outweigh resultant profits

EXERCISE 2   *page* 95

1. Median is 1·3 children per family.   3. 169 cm; 172·2 cm

CHAPTER 11

EXERCISE 1   *page* 99

1. 70 cm²   2. 7·2 m²   3. 7·73 m²   4. 0·47 km²
5. 449·4 cm²   6. 34 cm; 10·8 m; 12·78 m; 25·77 km; 12·85 m
7. 22 m   8. 34·5 m²   9. 1970 m
10. 11·56% (correct to 2 decimal places)

EXERCISE 2   *page* 101

1. (a) 36 cm²   (b) $36\frac{1}{2}$ cm²   (c) $85\frac{1}{2}$ cm²
2. 187·5 cm²; 1:3   3. 77·4 m²; 258 turves; £7·74
4. 97·5 m²; 8·97, i.e. 9 pieces required, but this allows nothing for pattern
   matching; 4·92 *and* 4·05, i.e. 10 pieces, assuming perfection in hanging
5. 2·08 m² (2 dec. pl.)
6. 3·40 m² (the error, 0·0002, is negligible in area and in cost); £1·17

T.Y.A.—12

EXERCISE 3 *page* 105

1. 50·4 tonnes   2. 7 min 34 s   3. 4 mm   4. 42 tonnes
5. 7180 cm³; the second part is tricky as it depends on the way in which the box is constructed, but we have allowed for concealed joints, which need most wood, i.e. we have used the *external* measurements of the box, the cost then being 55p (nearest 1p)   6. £12·10

EXERCISE 4 *page* 109

1. (a) 27 cm²   (b) 4·5 cm²   2. 16 cm   3. 1·517 m²
4. 446 litres (nearest litre)   5. 2584 bricks   6. 407 kg

EXERCISE 5 *page* 113

1. 73   2. 962   3. 3016   4. 1·414
5. 4·102   6. 2·222   7. 14·83   8. 80·14
9. 0·6301   10. 0·2417   11. 0·086 95   12. 0·027 80

EXERCISE 6 *page* 116

1. (a) 29   (b) 12   (c) 60   (d) 53   2. 65
3. 7·616   4. 19·26   5. 519·9   6. 104·6
7. 1·122   8. 16·52 cm   9. 6·86 m
10. 2·8 m; 8·54 m²   11. £44·38   12. 259 m   13. 6·40 cm

CHAPTER 12

EXERCISE 1 *page* 123

1. (a) 0·7782   (b) 0·9638   (c) 1·4914   (d) 0·2148
(e) 0·8603   (f) 0·8116   (g) 1·9355   (h) 1·2945
(i) 0·6852   (j) 1·5106   (k) 2·5014   (l) 0·0203
(m) 5·2148   (n) 0·9955   (o) 3·7364   (p) 1·8511
2. log 6·5 = 0·81, log 3·7 = 0·57, log 8·3 = 0·92 ⎫ correct to 2 sig. fig.
0·42 = log 2·6, 0·59 = log 3·9, 0·20 = log 1·6 ⎭

EXERCISE 2 *page* 125

1. (a) 4·125   (b) 2·017   (c) 7·934   (d) 1·004
(e) 18·62   (f) 12·05   (g) 293·6   (h) 40 000
(i) 9·725   (j) 6·009   (k) 2622   (l) 324 300

EXERCISE 3 *page* 126

1. 31·85   2. 125·9   3. 888·8   4. 34·53   5. 7·006
6. 91 180   7. 3·930   8. 1·052   9. 5·194   10. 183·1
11. 3·753   12. 1539   13. 15·46   14. 1·789   15. 14·66
16. 4·990   17. 1·539   18. 10 570   19. 1304   20. 8·474

EXERCISE 4 *page* 127

1. (a) 0   (b) $\bar{1}$ (c) 4   (d) $\bar{4}$ (e) $\bar{2}$ (f) $\bar{5}$ (g) $\bar{2}$
2. (a) $\bar{1}$·3412   (b) $\bar{2}$·7993   (c) $\bar{4}$·3381   (d) $\bar{4}$·9789
(e) $\bar{2}$·9586   (f) $\bar{1}$·9364   (g) 7·7924
3. (a) 441·6   (b) 0·044 16   (c) 0·9632   (d) 0·000 533 3
(e) 7 943 000   (f) 0·008 489   (g) 0·1023   (h) 0·000 000 587

EXERCISE 5 *page* 130

1. 1·769	2. 0·3685	3. 1·551	4. 0·001 174
5. 0·8798	6. 0·3122	7. 0·089 58	8. 0·009 811
9. 0·3533	10. 0·004 270	11. 2·179	12. 190·4
13. 80·04	14. 0·1607	15. 0·5347	16. 0·7711

EXERCISE 6 *page* 132

1. —0·105   2. 1·66   3. 2·54   4. (i) 20·1,   (ii) 793, (iii) 170
5. 28·8   6. (i) 137·(5),   (ii) 6·61   7. 9·33   8. 2·51

CHAPTER 13

EXERCISE 1 *page* 137

1. (a) 88 cm	(b) 66 cm	(c) 25·14 cm	(d) 4·714 m
2. (a) 13·9 cm²	(b) 1·86 m²	(c) 594 cm²	(d) 19·6 cm²
3. 7·90 cm; 24·8 cm	4. 24·2 cm	5. 5·01 m	6. 37 turns
7. 482 times	8. 65·9 times		

EXERCISE 2 *page* 142

1. 9·9 m³	2. 35·4 cm	3. 17 600 cm³	4. 155 kg
5. 644 cm²	6. 370 litres	7. 1·5 cm³	8. 5 min 14 s

EXERCISE 3 *page* 146

1. (a) 183 cm³   (b) 754 cm³   (c) 22·4 m³
2. (a) 178 cm²   (b) 133 cm²   (c) 28·1 m²
3. 127 cm²   4. 4·52 m³   5. 7850 tonnes
6. 1⁷⁄₉ cm exactly ≏ 1·78 cm (Note that π cancels out.)

EXERCISE 4 *page* 148

1. (a) 180 cm³; 154 cm²   (b) 113 cm³; 113 cm²
(c) 4·19 m²; 12·6 m²   (d) 2·5 m³; 8·91 m²
2. 98·5 cm   3. 8·92 cm   4. 2530 cm²   5. 0·496 cm
6. 39·6 m³   7. 82·6 cm²; 488 cm³ (We obtained a complete cone and
a hemisphere.)

CHAPTER 14

EXERCISE 1 *page* 152

1. £72	2. £85·50	3. £90·65	4. £1408	5. £17·81
6. £5·97	7. £128·16	8. £500	9. 8⅓%	10. £233·33
11. 8 years	12. £1600	13. £4·20	14. £4160	

EXERCISE 2 *page* 155

1. £65·60   2. £218·51   3. £66·11   4. £138·03   5. £90·77
6. £106·22   7. (a) 36p   (b) £25·41

EXERCISE 3   *page* 158

1. £652·80   2. £1954·67   3. £1122·45   4. £606·63   5. £1068
6. £2821   7. £568   8. £7420   9. 4·6%
10. (*a*) 17·7, (*b*) 9·01   11. 10%

CHAPTER 15

EXERCISE 1   *page* 162

1. £99·84   2. 57p   3. 65·47p   4. £172 185·60

EXERCISE 2   *page* 167

1. £1246·78   2. £1600·72   3. £110·82   4. £430·10
5. £2720 (nearest £10)

CHAPTER 16

EXERCISE 1   *Page* 175

1. £8000 stock; £280   2. 11·90%; £37·50   3. The latter; but only just!
4. £910; £41·25; 4·53%
5. 46op; without further information available, three possibilities come to
   mind; *either* capital appreciation *or* bonus share issue likely *or* increased
   dividend because of good earnings (or some combination of these).
6. 12½%   7. £5056; £41·48 less   8. 480   9. 30·4%; £107·80

EXERCISE 2   *page* 180

1. £1599·04   2. £547·22; £27·72   3. £24·13   4. £2093·12
5. 287 shares; £22·07; £7·48 (Minimal brokerage £7; see p. 175)

EXERCISE 3   *page* 182

1. £463·32   2. £9·90   3. £147·77

EXERCISE 4   *page* 183

1. £2555·22   2. 398 units; £5·13; £151·24   3. £44·28; £8·40
4. £8530 stock; £469·15

CHAPTER 17

EXERCISE 1   *page* 189

1. £26·46   2. £3·11; 5·07%   3. £54·56; £1·50   4. £5242·79
5. £68·25

CHAPTER 18

EXERCISE 1   *page* 196

1. (*a*) 1110   (*b*) 100010   (*c*) 11001101   (*d*) 1000100110
   (*e*) 100101001011

2. (a) 6   (b) 3   (c) 21   (d) 38   (e) 459
3. (a) 100001; 33   (b) 1000111; 71
   (c) 110111101; 445   (d) 100000100010; 4162
4. 1100110   5. 119   6. 11110
7. (a) 11001111; 207   (b) 11100101100; 1836
   (c) 111000111000; 3640   (d) 10001100011100; 8988
   (e) 1011100011001010; 47 306
8. 1111000 r. 1101; $120\frac{13}{18}$ 9. 1011 r. 1100; $11\frac{12}{19}$

LOGARITHMS
AND ANTI-LOGARITHMS

LOGARITHMS of numbers 1000 to 5499

	0	1	2	3	4	5	6	7	8	9	1	2	3	4	5	6	7	8	9
10	0000	0043	0086	0128	0170	0212	0253	0294	0334	0374	4	8	12	17	21	25	29	33	37
11	0414	0453	0492	0531	0569	0607	0645	0682	0719	0755	4	8	11	15	19	23	26	30	34
12	0792	0828	0864	0899	0934	0969	1004	1038	1072	1106	3	7	10	14	17	21	24	28	31
13	1139	1173	1206	1239	1271	1303	1335	1367	1399	1430	3	6	10	13	16	19	23	26	29
14	1461	1492	1523	1553	1584	1614	1644	1673	1703	1732	3	6	9	12	15	18	21	24	27
15	1761	1790	1818	1847	1875	1903	1931	1959	1987	2014	3	6	8	11	14	17	20	22	25
16	2041	2068	2095	2122	2148	2175	2201	2227	2253	2279	3	5	8	11	13	16	18	21	24
17	2304	2330	2355	2380	2405	2430	2455	2480	2504	2529	2	5	7	10	12	15	17	20	22
18	2553	2577	2601	2625	2648	2672	2695	2718	2742	2765	2	5	7	9	12	14	16	19	21
19	2788	2810	2833	2856	2878	2900	2923	2945	2967	2989	2	4	7	9	11	13	16	18	20
20	3010	3032	3054	3075	3096	3118	3139	3160	3181	3201	2	4	6	8	11	13	15	17	19
21	3222	3243	3263	3284	3304	3324	3345	3365	3385	3404	2	4	6	8	10	12	14	16	18
22	3424	3444	3464	3483	3502	3522	3541	3560	3579	3598	2	4	6	8	10	12	14	15	17
23	3617	3636	3655	3674	3692	3711	3729	3747	3766	3784	2	4	6	7	9	11	13	15	17
24	3802	3820	3838	3856	3874	3892	3909	3927	3945	3962	2	4	5	7	9	11	12	14	16
25	3979	3997	4014	4031	4048	4065	4082	4099	4116	4133	2	3	5	7	9	10	12	14	15
26	4150	4166	4183	4200	4216	4232	4249	4265	4281	4298	2	3	5	7	8	10	11	13	15
27	4314	4330	4346	4362	4378	4393	4409	4425	4440	4456	2	3	5	6	8	9	11	13	14
28	4472	4487	4502	4518	4533	4548	4564	4579	4594	4609	2	3	5	6	8	9	11	12	14
29	4624	4639	4654	4669	4683	4698	4713	4728	4742	4757	1	3	4	6	7	9	10	12	13
30	4771	4786	4800	4814	4829	4843	4857	4871	4886	4900	1	3	4	6	7	9	10	11	13
31	4914	4928	4942	4955	4969	4983	4997	5011	5024	5038	1	3	4	5	7	8	10	11	12
32	5051	5065	5079	5092	5105	5119	5132	5145	5159	5172	1	3	4	5	7	8	9	11	12
33	5185	5198	5211	5224	5237	5250	5263	5276	5289	5302	1	3	4	5	6	8	9	10	12
34	5315	5328	5340	5353	5366	5378	5391	5403	5416	5428	1	3	4	5	6	8	9	10	11
35	5441	5453	5465	5478	5490	5502	5514	5527	5539	5551	1	2	4	5	6	7	9	10	11
36	5563	5575	5587	5599	5611	5623	5635	5647	5658	5670	1	2	4	5	6	7	8	10	11
37	5632	5694	5705	5717	5729	5740	5752	5763	5775	5786	1	2	3	5	6	7	8	9	10
38	5798	5809	5821	5832	5843	5855	5866	5877	5888	5899	1	2	3	5	6	7	8	9	10
39	5911	5922	5933	5944	5955	5966	5977	5988	5999	6010	1	2	3	4	5	7	8	9	10
40	6021	6031	6042	6053	6064	6075	6085	6096	6107	6117	1	2	3	4	5	6	7	9	10
41	6128	6138	6149	6160	6170	6180	6191	6201	6212	6222	1	2	3	4	5	6	7	8	9
42	6232	6243	6253	6263	6274	6284	6294	6304	6314	6325	1	2	3	4	5	6	7	8	9
43	6335	6345	6355	6365	6375	6385	6395	6405	6415	6425	1	2	3	4	5	6	7	8	9
44	6435	6444	6454	6464	6474	6484	6493	6503	6513	6522	1	2	3	4	5	6	7	8	9
45	6532	6542	6551	6561	6571	6580	6590	6599	6609	6618	1	2	3	4	5	6	7	8	9
46	6628	6637	6646	6656	6665	6675	6684	6693	6702	6712	1	2	3	4	5	6	7	7	8
47	6721	6730	6739	6749	6758	6767	6776	6785	6794	6803	1	2	3	4	5	5	6	7	8
48	6812	6821	6830	6839	6848	6857	6866	6875	6884	6893	1	2	3	4	4	5	6	7	8
49	6902	6911	6920	6928	6937	6946	6955	6964	6972	6981	1	2	3	4	4	5	6	7	8
50	6990	6998	7007	7016	7024	7033	7042	7050	7059	7067	1	2	3	3	4	5	6	7	8
51	7076	7084	7093	7101	7110	7118	7126	7135	7143	7152	1	2	3	3	4	5	6	7	8
52	7160	7168	7177	7185	7193	7202	7210	7218	7226	7235	1	2	2	3	4	5	6	7	7
53	7243	7251	7259	7267	7275	7284	7292	7300	7308	7316	1	2	2	3	4	5	6	6	7
54	7324	7332	7340	7348	7356	7364	7372	7380	7388	7396	1	2	2	3	4	5	6	6	7
	0	1	2	3	4	5	6	7	8	9	1	2	3	4	5	6	7	8	9

Proportional Parts

	0	1	2	3	4	5	6	7	8	9	1	2	3	4	5	6	7	8	9
55	7404	7412	7419	7427	7435	7443	7451	7459	7466	7474	1	2	2	3	4	5	5	6	7
56	7482	7490	7497	7505	7513	7520	7528	7536	7543	7551	1	2	2	3	4	5	5	6	7
57	7559	7566	7574	7582	7589	7597	7604	7612	7619	7627	1	2	2	3	4	5	5	6	7
58	7634	7642	7649	7657	7664	7672	7679	7686	7694	7701	1	1	2	3	4	4	5	6	7
59	7709	7716	7723	7731	7738	7745	7752	7760	7767	7774	1	1	2	3	4	4	5	6	7
60	7782	7789	7796	7803	7810	7818	7825	7832	7839	7846	1	1	2	3	4	4	5	6	6
61	7853	7860	7868	7875	7882	7889	7896	7903	7910	7917	1	1	2	3	4	4	5	6	6
62	7924	7931	7938	7945	7952	7959	7966	7973	7980	7987	1	1	2	3	4	4	5	6	6
63	7993	8000	8007	8014	8021	8028	8035	8041	8048	8055	1	1	2	3	3	4	5	6	6
64	8062	8069	8075	8082	8089	8096	8102	8109	8116	8122	1	1	2	3	3	4	5	5	6
65	8129	8136	8142	8149	8156	8162	8169	8176	8182	8189	1	1	2	3	3	4	5	5	6
66	8195	8202	8209	8215	8222	8228	8235	8241	8248	8254	1	1	2	3	3	4	5	5	6
67	8261	8267	8274	8280	8287	8293	8299	8306	8312	8319	1	1	2	3	3	4	4	5	6
68	8325	8331	8338	8344	8351	8357	8363	8370	8376	8382	1	1	2	3	3	4	4	5	6
69	8388	8395	8401	8407	8414	8420	8426	8432	8439	8445	1	1	2	3	3	4	4	5	6
70	8451	8457	8463	8470	8476	8482	8488	8494	8500	8506	1	1	2	2	3	4	4	5	6
71	8513	8519	8525	8531	8537	8543	8549	8555	8561	8567	1	1	2	2	3	4	4	5	5
72	8573	8579	8585	8591	8597	8603	8609	8615	8621	8627	1	1	2	2	3	4	4	5	5
73	8633	8639	8645	8651	8657	8663	8669	8675	8681	8686	1	1	2	2	3	4	4	5	5
74	8692	8698	8704	8710	8716	8722	8727	8733	8739	8745	1	1	2	2	3	4	4	5	5
75	8751	8756	8762	8768	8774	8779	8785	8791	8797	8802	1	1	2	2	3	3	4	5	5
76	8808	8814	8820	8825	8831	8837	8842	8848	8854	8859	1	1	2	2	3	3	4	5	5
77	8865	8871	8876	8882	8887	8893	8899	8904	8910	8915	1	1	2	2	3	3	4	4	5
78	8921	8927	8932	8938	8943	8949	8954	8960	8965	8971	1	1	2	2	3	3	4	4	5
79	8976	8982	8987	8993	8998	9004	9009	9015	9020	9025	1	1	2	2	3	3	4	4	5
80	9031	9036	9042	9047	9053	9058	9063	9069	9074	9079	1	1	2	2	3	3	4	4	5
81	9085	9090	9096	9101	9106	9112	9117	9122	9128	9133	1	1	2	2	3	3	4	4	5
82	9138	9143	9149	9154	9159	9165	9170	9175	9180	9186	1	1	2	2	3	3	4	4	5
83	9191	9196	9201	9206	9212	9217	9222	9227	9232	9238	1	1	2	2	3	3	4	4	5
84	9243	9248	9253	9258	9263	9269	9274	9279	9284	9289	1	1	2	2	3	3	4	4	5
85	9294	9299	9304	9309	9315	9320	9325	9330	9335	9340	1	1	2	2	3	3	4	4	5
86	9345	9350	9355	9360	9365	9370	9375	9380	9385	9390	1	1	2	2	3	3	4	4	5
87	9395	9400	9405	9410	9415	9420	9425	9430	9435	9440	0	1	1	2	2	3	3	4	4
88	9445	9450	9455	9460	9465	9469	9474	9479	9484	9489	0	1	1	2	2	3	3	4	4
89	9494	9499	9504	9509	9513	9518	9523	9528	9533	9538	0	1	1	2	2	3	3	4	4
90	9542	9547	9552	9557	9562	9566	9571	9576	9581	9586	0	1	1	2	2	3	3	4	4
91	9590	9595	9600	9605	9609	9614	9619	9624	9628	9633	0	1	1	2	2	3	3	4	4
92	9638	9643	9647	9652	9657	9661	9666	9671	9675	9680	0	1	1	2	2	3	3	4	4
93	9685	9689	9694	9699	9703	9708	9713	9717	9722	9727	0	1	1	2	2	3	3	4	4
94	9731	9736	9741	9745	9750	9754	9759	9764	9768	9773	0	1	1	2	2	3	3	4	4
95	9777	9782	9786	9791	9795	9800	9805	9809	9814	9818	0	1	1	2	2	3	3	4	4
96	9823	9827	9832	9836	9841	9845	9850	9854	9859	9863	0	1	1	2	2	3	3	4	4
97	9868	9872	9877	9881	9886	9890	9894	9899	9903	9908	0	1	1	2	2	3	3	4	4
98	9912	9917	9921	9926	9930	9934	9939	9943	9948	9952	0	1	1	2	2	3	3	4	4
99	9956	9961	9965	9969	9974	9978	9983	9987	9991	9996	0	1	1	2	2	3	3	4	4
	0	1	2	3	4	5	6	7	8	9	1	2	3	4	5	6	7	8	9

ANTI-LOGARITHMS

	0	1	2	3	4	5	6	7	8	9	1	2	3	4	5	6	7	8	9
·00	1000	1002	1005	1007	1009	1012	1014	1016	1019	1021	0	0	1	1	1	1	2	2	2
·01	1023	1026	1028	1030	1033	1035	1038	1040	1042	1045	0	0	1	1	1	1	2	2	2
·02	1047	1050	1052	1054	1057	1059	1062	1064	1067	1069	0	0	1	1	1	1	2	2	2
·03	1072	1074	1076	1079	1081	1084	1086	1089	1091	1094	0	0	1	1	1	1	2	2	2
·04	1096	1099	1102	1104	1107	1109	1112	1114	1117	1119	0	1	1	1	1	2	2	2	2
·05	1122	1125	1127	1130	1132	1135	1138	1140	1143	1146	0	1	1	1	1	2	2	2	2
·06	1148	1151	1153	1156	1159	1161	1164	1167	1169	1172	0	1	1	1	1	2	2	2	2
·07	1175	1178	1180	1183	1186	1189	1191	1194	1197	1199	0	1	1	1	1	2	2	2	2
·08	1202	1205	1208	1211	1213	1216	1219	1222	1225	1227	0	1	1	1	1	2	2	2	3
·09	1230	1233	1236	1239	1242	1245	1247	1250	1253	1256	0	1	1	1	1	2	2	2	3
·10	1259	1262	1265	1268	1271	1274	1276	1279	1282	1285	0	1	1	1	1	2	2	2	3
·11	1288	1291	1294	1297	1300	1303	1306	1309	1312	1315	0	1	1	1	2	2	2	2	3
·12	1318	1321	1324	1327	1330	1334	1337	1340	1343	1346	0	1	1	1	2	2	2	3	3
·13	1349	1352	1355	1358	1361	1365	1368	1371	1374	1377	0	1	1	1	2	2	2	3	3
·14	1380	1384	1387	1390	1393	1396	1400	1403	1406	1409	0	1	1	1	2	2	2	3	3
·15	1413	1416	1419	1422	1426	1429	1432	1435	1439	1442	0	1	1	1	2	2	2	3	3
·16	1445	1449	1452	1455	1459	1462	1466	1469	1472	1476	0	1	1	1	2	2	2	3	3
·17	1479	1483	1486	1489	1493	1496	1500	1503	1507	1510	0	1	1	1	2	2	2	3	3
·18	1514	1517	1521	1524	1528	1531	1535	1538	1542	1545	0	1	1	1	2	2	2	3	3
·19	1549	1552	1556	1560	1563	1567	1570	1574	1578	1581	0	1	1	1	2	2	3	3	3
·20	1585	1589	1592	1596	1600	1603	1607	1611	1614	1618	0	1	1	1	2	2	3	3	3
·21	1622	1626	1629	1633	1637	1641	1644	1648	1652	1656	0	1	1	2	2	2	3	3	3
·22	1660	1663	1667	1671	1675	1679	1683	1687	1690	1694	0	1	1	2	2	2	3	3	3
·23	1698	1702	1706	1710	1714	1718	1722	1726	1730	1734	0	1	1	2	2	2	3	3	4
·24	1738	1742	1746	1750	1754	1758	1762	1766	1770	1774	0	1	1	2	2	2	3	3	4
·25	1778	1782	1786	1791	1795	1799	1803	1807	1811	1816	0	1	1	2	2	3	3	3	4
·26	1820	1824	1828	1832	1837	1841	1845	1849	1854	1858	0	1	1	2	2	3	3	3	4
·27	1862	1866	1871	1875	1879	1884	1888	1892	1897	1901	0	1	1	2	2	3	3	3	4
·28	1905	1910	1914	1919	1923	1928	1932	1936	1941	1945	0	1	1	2	2	3	3	4	4
·29	1950	1954	1959	1963	1968	1972	1977	1982	1986	1991	0	1	1	2	2	3	3	4	4
·30	1995	2000	2004	2009	2014	2018	2023	2028	2032	2037	0	1	1	2	2	3	3	4	4
·31	2042	2046	2051	2056	2061	2065	2070	2075	2080	2084	0	1	1	2	2	3	3	4	4
·32	2089	2094	2099	2104	2109	2113	2118	2123	2128	2133	0	1	1	2	2	3	3	4	4
·33	2138	2143	2148	2153	2158	2163	2168	2173	2178	2183	0	1	1	2	2	3	3	4	4
·34	2188	2193	2198	2203	2208	2213	2218	2223	2228	2234	1	1	2	2	3	3	4	4	5
·35	2239	2244	2249	2254	2259	2265	2270	2275	2280	2286	1	1	2	2	3	3	4	4	5
·36	2291	2296	2301	2307	2312	2317	2323	2328	2333	2339	1	1	2	2	3	3	4	4	5
·37	2344	2350	2355	2360	2366	2371	2377	2382	2388	2393	1	1	2	2	3	3	4	4	5
·38	2399	2404	2410	2415	2421	2427	2432	2438	2443	2449	1	1	2	2	3	3	4	4	5
·39	2455	2460	2466	2472	2477	2483	2489	2495	2500	2506	1	1	2	2	3	3	4	5	5
·40	2512	2518	2523	2529	2535	2541	2547	2553	2559	2564	1	1	2	2	3	3	4	5	5
·41	2570	2576	2582	2588	2594	2600	2606	2612	2618	2624	1	1	2	2	3	4	4	5	5
·42	2630	2636	2642	2648	2655	2661	2667	2673	2679	2685	1	1	2	2	3	4	4	5	6
·43	2692	2698	2704	2710	2716	2723	2729	2735	2742	2748	1	1	2	2	3	4	4	5	6
·44	2754	2761	2767	2773	2780	2786	2793	2799	2805	2812	1	1	2	3	3	4	4	5	6
·45	2818	2825	2831	2838	2844	2851	2858	2864	2871	2877	1	1	2	3	3	4	5	5	6
·46	2884	2891	2897	2904	2911	2917	2924	2931	2938	2944	1	1	2	3	3	4	5	5	6
·47	2951	2958	2965	2972	2979	2985	2992	2999	3006	3013	1	1	2	3	3	4	5	6	6
·48	3020	3027	3034	3041	3048	3055	3062	3069	3076	3083	1	1	2	3	4	4	5	6	6
·49	3090	3097	3105	3112	3119	3126	3133	3141	3148	3155	1	1	2	3	4	4	5	6	7
	0	1	2	3	4	5	6	7	8	9	1	2	3	4	5	6	7	8	9

ANTI-LOGARITHMS

	0	1	2	3	4	5	6	7	8	9	1	2	3	4	5	6	7	8	9
·50	3162	3170	3177	3184	3192	3199	3206	3214	3221	3228	1	1	2	3	4	4	5	6	7
·51	3236	3243	3251	3258	3266	3273	3281	3289	3296	3304	1	2	2	3	4	5	5	6	7
·52	3311	3319	3327	3334	3342	3350	3357	3365	3373	3381	1	2	2	3	4	5	5	6	7
·53	3388	3396	3404	3412	3420	3428	3436	3443	3451	3459	1	2	2	3	4	5	6	6	7
·54	3467	3475	3483	3491	3499	3508	3516	3524	3532	3540	1	2	2	3	4	5	6	6	7
·55	3548	3556	3565	3573	3581	3589	3597	3606	3614	3622	1	2	2	3	4	5	6	7	7
·56	3631	3639	3648	3656	3664	3673	3681	3690	3698	3707	1	2	3	3	4	5	6	7	8
·57	3715	3724	3733	3741	3750	3758	3767	3776	3784	3793	1	2	3	3	4	5	6	7	8
·58	3802	3811	3819	3828	3837	3846	3855	3864	3873	3882	1	2	3	4	4	5	6	7	8
·59	3890	3899	3908	3917	3926	3936	3945	3954	3963	3972	1	2	3	4	5	5	6	7	8
·60	3981	3990	3999	4009	4018	4027	4036	4046	4055	4064	1	2	3	4	5	6	6	7	8
·61	4074	4083	4093	4102	4111	4121	4130	4140	4150	4159	1	2	3	4	5	6	7	8	9
·62	4169	4178	4188	4198	4207	4217	4227	4236	4246	4256	1	2	3	4	5	6	7	8	9
·63	4266	4276	4285	4295	4305	4315	4325	4335	4345	4355	1	2	3	4	5	6	7	8	9
·64	4365	4375	4385	4395	4406	4416	4426	4436	4446	4457	1	2	3	4	5	6	7	8	9
·65	4467	4477	4487	4498	4508	4519	4529	4539	4550	4560	1	2	3	4	5	6	7	8	9
·66	4571	4581	4592	4603	4613	4624	4634	4645	4656	4667	1	2	3	4	5	6	7	8	10
·67	4677	4688	4699	4710	4721	4732	4742	4753	4764	4775	1	2	3	4	5	7	8	9	10
·68	4786	4797	4808	4819	4831	4842	4853	4864	4875	4887	1	2	3	4	6	7	8	9	10
·69	4898	4909	4920	4932	4943	4955	4966	4977	4989	5000	1	2	3	5	6	7	8	9	10
·70	5012	5023	5035	5047	5058	5070	5082	5093	5105	5117	1	2	4	5	6	7	8	9	11
·71	5129	5140	5152	5164	5176	5188	5200	5212	5224	5236	1	2	4	5	6	7	8	10	11
·72	5248	5260	5272	5284	5297	5309	5321	5333	5346	5358	1	2	4	5	6	7	9	10	11
·73	5370	5383	5395	5408	5420	5433	5445	5458	5470	5483	1	3	4	5	6	8	9	10	11
·74	5495	5508	5521	5534	5546	5559	5572	5585	5598	5610	1	3	4	5	6	8	9	10	12
·75	5623	5636	5649	5662	5675	5689	5702	5715	5728	5741	1	3	4	5	7	8	9	10	12
·76	5754	5768	5781	5794	5808	5821	5834	5848	5861	5875	1	3	4	5	7	8	9	11	12
·77	5888	5902	5916	5929	5943	5957	5970	5984	5998	6012	1	3	4	6	7	8	10	11	12
·78	6026	6039	6053	6067	6081	6095	6109	6124	6138	6152	1	3	4	6	7	8	10	11	13
·79	6166	6180	6194	6209	6223	6237	6252	6266	6281	6295	1	3	4	6	7	9	10	12	13
·80	6310	6324	6339	6353	6368	6383	6397	6412	6427	6442	1	3	4	6	7	9	10	12	13
·81	6457	6471	6486	6501	6516	6531	6546	6561	6577	6592	2	3	5	6	8	9	11	12	14
·82	6607	6622	6637	6653	6668	6683	6699	6714	6730	6745	2	3	5	6	8	9	11	12	14
·83	6761	6776	6792	6808	6823	6839	6855	6871	6887	6902	2	3	5	6	8	9	11	13	14
·84	6918	6934	6950	6966	6982	6998	7015	7031	7047	7063	2	3	5	6	8	10	11	13	14
·85	7079	7096	7112	7129	7145	7161	7178	7194	7211	7228	2	3	5	7	8	10	12	13	15
·86	7244	7261	7278	7295	7311	7328	7345	7362	7379	7396	2	3	5	7	8	10	12	14	15
·87	7413	7430	7447	7464	7482	7499	7516	7534	7551	7568	2	3	5	7	9	10	12	14	16
·88	7586	7603	7621	7638	7656	7674	7691	7709	7727	7745	2	4	5	7	9	11	12	14	16
·89	7762	7780	7798	7816	7834	7852	7870	7889	7907	7925	2	4	5	7	9	11	13	14	16
·90	7943	7962	7980	7998	8017	8035	8054	8072	8091	8110	2	4	6	7	9	11	13	15	17
·91	8128	8147	8166	8185	8204	8222	8241	8260	8279	8299	2	4	6	8	10	11	13	15	17
·92	8318	8337	8356	8375	8395	8414	8433	8453	8472	8492	2	4	6	8	10	12	14	15	17
·93	8511	8531	8551	8570	8590	8610	8630	8650	8670	8690	2	4	6	8	10	12	14	16	18
·94	8710	8730	8750	8770	8790	8810	8831	8851	8872	8892	2	4	6	8	10	12	14	16	18
·95	8913	8933	8954	8974	8995	9016	9036	9057	9078	9099	2	4	6	8	10	12	14	17	19
·96	9120	9141	9162	9183	9204	9226	9247	9268	9290	9311	2	4	6	9	11	13	15	17	19
·97	9333	9354	9376	9397	9419	9441	9462	9484	9506	9528	2	4	7	9	11	13	15	17	20
·98	9550	9572	9594	9616	9638	9661	9683	9705	9727	9750	2	4	7	9	11	13	16	18	20
·99	9772	9795	9817	9840	9863	9886	9908	9931	9954	9977	2	5	7	9	11	14	16	18	21
	0	1	2	3	4	5	6	7	8	9	1	2	3	4	5	6	7	8	9

INDEX

Abacus, 1
Answers, 201
Area, 47
  circle, 134
  rectangle, 97
  trapezium, 107
  triangle, 105
Arithmetical graphs, 87
Assurance, 164
Average, 73

Binary scale, 192
Box, 104
Brackets, 16
Brokerage, 176

Capacity, 50
Characteristic, 135, 141
Circle, 134
  area, 135
  chord, 134
  circumference, 134
  diameter, 134
  radius, 134
Coinage, 54
Compound interest, 153
  by logarithms, 157
  formula, 155
Computing machine, 191
Cone, 143
  surface, 144
  volume, 144

Decimalization of money, 57
Decimals, 35
  addition and subtraction, 36
  multiplication and division, 37
Digital computer, 191
Discount, 83

Factors, 19
  division by, 21
  prime, 20
  tests for, 19
Four rules (of number), 9

Fractions, 24
  addition and subtraction, 27
  division, 30
  mixed signs, 31
  multiplication, 29
  problems, 33
Frequency distribution, 92

Graphs, arithmetical, 87

Highest common factor, 22
Histogram, 92

Income Tax, 162
  on investment, 181
  Schedule E, 164
Indices, 118
Insurance, 164
  life assurance, 164
  property, 166
Interest, 150
  compound, 153
  simple, 150
Investment, 169

Length, 45
Loan, repayment of, 158
Logarithms, 120
  antilogarithms, 124
  four rules of, 123
  graphical considerations, 120
  negative characteristics, 126
  problems and formulae, 130
  tables, 213–216
  use of tables, 120
Lowest common multiple, 22

Mantissa, 120
Mass, 50
Mean, 96
Median, 95
Mensuration, 97
Metric system, 44
Mixtures, 84

Money, 54
    addition, 58
    division, 60
    multiplication, 60
    practice, 63
    subtraction, 60
    unitary method, 57

Numbers, 8
    addition, 9
    brackets, etc., 16
    division, 14
    multiplication, 12
    subtraction, 11
Numerals, 1

Ogive, 95

Percentage, 76
    problems, 81
    profit, 78
Perimeter, 98
Plane figure, 99
Pipe, 140
π, value of, 135
Practice, 63
Principal, 150
Prism, 108
Proportion, 68
Proportional parts, 72

Quartile, 95

Rates, 160
    water, 162
Ratio, 66

Rectangle, 97
Right-angled triangle, 113

Simple interest, 150
    formula, 150
Sphere, 147
    surface, 147
    volume, 147
Square root, 110
Stamp duty, 176
Stock Exchange, 170
Stocks and shares, 171
Surface of,
    cone, 143
    cylinder, 137
    sphere, 147

Tables, xiii–xvi
    logarithm, 213–216
Tax, 160, 162
Trapezium, 107
Triangle, 105, 113

Unit trusts, 182
Unitary method, 64

Variation, 68
    inverse, 69
Volume, 50
    cone, 143
    cuboid, 102
    cylinder, 137
    prism, 108
    sphere, 147

Weight (see Mass)